THE
INTIMATE
UNIVERSE

THE INTIMATE UNIVERSE

HOW THE STARS ARE CLOSER THAN YOU THINK

MAREK KUKULA

Quercus  ROYAL
OBSERVATORY
GREENWICH

First published in Great Britain in 2015 by

Quercus Publishing Ltd
Carmelite House
50 Victoria Embankment
London EC4Y 0DZ

An Hachette UK company

Published in association with Royal Museums Greenwich,
the group name for the National Maritime Museum,
Royal Observatory Greenwich, Queen's House and *Cutty Sark*.

www.rmg.co.uk

Text © 2015 National Maritime Museum, Greenwich, London
Author: Marek Kukula

A CIP catalogue record for this book is available
from the British Library

HB ISBN 978 1 78206 789 4
TPB ISBN 978 1 78429 487 8
EBOOK ISBN 978 1 78206 790 0

Every effort has been made to contact copyright holders.
However, the publishers will be glad to rectify in future editions
any inadvertent omissions brought to their attention.

10 9 8 7 6 5 4 3 2 1

Typeset by Hewer Text UK Ltd, Edinburgh
Printed and bound in Great Britain by Clays Ltd, St Ives plc

For my father

Contents

Introduction

We live in an interconnected world. Our daily lives depend on networks of trade, transport and communication that extend across the planet. Food and raw materials are transported halfway around the globe, while local and national economies are inextricably bound into a global system, so much so that its upswings and crashes affect us all. Vast quantities of information surge along the virtual highways of the internet, satellites beam global news into our homes, and individuals, nations and cultures interact with an ease that no one could have imagined just a few decades ago, leading to both conflict and cooperation in new and unpredictable ways.

While the human population of planet Earth is linked together as never before, this global web of connections is by no means limited to the affairs of the human race. As we have learned more and more about the workings of our planet's geology, atmosphere, oceans and biosphere, we have come to understand that these too are joined together in a single global system that has been shaping our world ever since it formed 4.5 billion years ago.

As we humans have discovered to our cost, the Earth's natural systems do not respect national boundaries. In the 1970s, sulphur dioxide emissions from coal-fired power stations in Britain led to acid rain in Scandinavia, poisoning the native conifer forests and damaging the local timber industry. In 1986, radioactive particles from the burning Chernobyl nuclear reactor in the

Soviet Union rained down on the mountains of North Wales, leading to a ban on human consumption of livestock reared on the contaminated pastures. Today, global warming primarily caused by the carbon dioxide emissions of industrialized nations is recognized as a problem that affects the entire world. On an interconnected planet no nation is immune from the actions of its neighbours.

Natural events in one place can also have profound consequences on the other side of the planet. Sandstorms in the Sahara blow desert dust high into the atmosphere, where it is carried across the Atlantic to fertilize the rainforests of the Amazon Basin. El Niño events warm the surface waters of the Pacific Ocean, triggering changes in rainfall and temperature around the globe, with floods in California, droughts in Australia and colder winters in Northern Europe. Even the day-to-day vagaries of the world's weather are linked together by a bewilderingly complex web of connections: as the old cliché reminds us, the beat of a butterfly's wings in Brazil might lead, via a global chain of atmospheric dominoes, to a thunderstorm in Birmingham.

But Earth itself is not a closed system, cut off from the rest of the universe. Our environment does not stop at the top of the atmosphere and there is no invisible barrier there to seal our planet into an impenetrable bubble. On the contrary, planet Earth is constantly interacting with its surroundings, and extraterrestrial influences continue to shape our world in every conceivable way. From the sunlight that warms us, driving our weather and powering our food chains, to the gravitational forces that control the tides and keep the Earth on its seasonal track, objects in space exert a profound and immediate influence on our daily lives. The connections go much deeper than that: every molecule, atom and subatomic particle here on Earth can trace its origins out into the wider cosmos, each with an astonishing story to tell, involving comets and nebulae, titanic collisions and exploding stars, and stretching out into the depths of the galaxy and back to the origins of the universe itself. To understand how our planet came to be the way it is, and how its life-sustaining geology, climate and biosphere operate today, we need to consider the Earth as a component in a much larger cosmic system.

Even some of the strangest and most extreme types of astronomical phenomena have closer connections to our daily lives than we might suppose. Exotic objects like neutron stars and supernova remnants have played a crucial role in the creation of the habitable conditions that we enjoy here on Earth, while the alien geologies and atmospheres of our neighbouring planets and moons provide us with profound insights into the workings of our own world.

Astronomy has also taught us to see our immediate surroundings in a strange new light. One of the greatest achievements of modern cosmology has been the discovery that the ordinary matter that forms our bodies, the Earth, stars, nebulae and galaxies – everything in fact that our telescopes can see – is only 5 per cent of the total content of the universe. The rest seems to be made up of two intriguing but mysterious quantities: dark matter, which holds the galaxies together and accounts for 27 per cent of the cosmic total, and the even more enigmatic dark energy, which provides the remaining 68 per cent and which seems to be causing the universe to expand faster and faster. We currently have no firm idea of what dark matter and dark energy consist of – in other words an astonishing 95 per cent of the stuff of the universe remains a mystery. And yet dark matter and dark energy are all around us – not just out there in the distant reaches of interstellar space but here in the room with us now. Between them, they have played a profound role in shaping the universe of galaxies, stars and planets, helping to create the conditions in which life can exist, but although they have been our constant companions for the whole of human history, it is only in the last few decades that we have realized they exist.

This book is all about the surprisingly intimate connections that we have with the universe beyond our home world. Over the coming chapters we will explore the cosmic origins of the familiar substances that make up the world around us: the Earth's rocks, air and oceans, and even our own bodies. We will discover how the conditions that we take for granted here on Earth are the result of the cataclysmic forces that shaped stars, solar systems and galaxies and how our closest neighbours in space, the Sun and Moon, continue to exert a profound influence on every aspect of our planet. We will see how the Earth's ceaseless motions through space determine the cycles of days and years and how physics and astronomy enable us to peer into the future and glimpse the ultimate fate of the Earth, the Sun and the universe itself. In the process we will learn how this knowledge has affected human culture – and how our species is beginning to make its mark beyond our home planet.

Astronomy and space science may be concerned with the observation and exploration of distant planets, stars and galaxies but these objects are not divorced from our everyday concerns here on Earth. On the contrary, the origins of our planet, its current status and even its future fate can only be properly understood in the broader context of its cosmic environment. As we shall discover, everything we learn about these remote objects helps us to see our own world in a clearer light.

Made of Starstuff

Where do we come from? It's a simple question, but one which goes right to the heart of how we think about ourselves, our relationship to the planet and to the rest of the universe. We all have our own personal stories, woven together out of the events, places and people that have influenced us and made us who we are; but, as well as shaping our personalities and memories, these life stories also leave their mark on the physical stuff of which we're made. Throughout our lives, every time we eat, drink and inhale, new atoms and molecules enter our bodies. Meanwhile, as we exhale and excrete, other atoms and molecules are removed and sent back out into the wider world. Many of the new arrivals only spend a brief time inside our bodies, taking part in the various biochemical processes that keep us alive before leaving again. Others become a more permanent part of us, being incorporated into the structures of our organs, blood and bones.

Our bodies are the sum of the trillions of different molecules that help to build and sustain them over a lifetime, but each atom and molecule also has its own life story – a story that stretches back way before our own lives, and which will extend far beyond us into the future: ultimately all of the atoms and molecules here on Earth have their origins somewhere far more alien than our own planet.

This direct connection between our bodies and the stars is astonishing enough, but the story of how the stuff that we're made of was created, and the immense journey through space and time that it took before it reached us, is one of the most extraordinary things that astronomy has taught us.

It is relatively easy to find out where a manufactured object like a fridge or a car was made, even if this was in a factory halfway across the world from where we bought it. But this place is very likely just where the object was assembled. If we investigated a little further, we could probably find out where each of the individual components was sourced from, and by digging deeper still we could even trace the origins of the raw materials from which they were built. Eventually, all these trails would lead back to the mines from which various minerals were extracted, or the fields, forests and seas where plant- and animal-based materials were grown and harvested.

By analysing bones, teeth and hair, archaeologists and forensic scientists can perform a similar piece of detective work on a human being. A record of where they were born and grew up, what they ate, drank and breathed, and even when and where they died, is written in the subtle chemical signatures inside them. These correspond to the differing chemical compositions of the environments in which they spent their lives breathing, eating, drinking and simply coming into contact with their surroundings.

The Earth's great cycles of water, carbon, nitrogen and other substances ensure that there is constant traffic between the atmosphere, rocks, oceans and biosphere of our planet. The molecules of water that we excrete might eventually find their way via sewers, streams and rivers back to the sea, only to evaporate, condense into clouds and fall as rain, perhaps to be consumed and, briefly, become part of a human body once again.

Carbon atoms experience a similar circulation between living things and their environment. The carbon dioxide that we exhale is the waste product of the process that powers our bodies. Deep inside our cells, carbon-rich molecules such as sugars combine with the oxygen we've breathed, releasing useful energy and producing carbon dioxide in the process. The carbon dioxide is then whisked away by our blood and exits via our lungs – but this is by no means the end of its story.

Some of this exhaled carbon dioxide may get sucked up by the leaves of nearby plants, which will use the energy from sunlight to combine it with water to form sugar molecules again. These sugars are intended to be the food supply for the plant itself but they may also be stolen by animals which eat the plant, quickly re-entering the food chain and perhaps ultimately finding themselves back inside a human being again. But if the plant dies and is buried, gradually being compressed by geological processes into coal, the captured carbon will be locked away in the ground and thus removed from the cycle for millions of years – at least until humans dig it up and burn it, releasing it back into the atmosphere as carbon dioxide once more.

Other molecules of carbon dioxide might dissolve in the ocean, forming first carbonic acid then carbonates, which react with dissolved calcium to form calcium carbonate – the material that makes up the shells of sea creatures and ultimately is deposited on the sea bed to form limestone, once again locking the carbon atoms away for millions of years.

Cycles such as these have been taking place in some form or other over most of the 4.5-billion-year history of the Earth, and they play a major role in maintaining our planet's habitability. One way or another, most of the atoms in our bodies have been taking part in the sagas of the planet's rocks, air and oceans for a lot longer than they have been inside living creatures, combining with each other in fleeting or long-lasting chemical alliances to form the vast array of molecules that we find here on Earth. But before the Earth had even formed, these atoms had already been in existence for millions or even billions of years. Based on the fraction of their time that they have spent here, it is fair to say that the cosy environs of Earth are not the natural habitat of most of the atoms that surround us. From biology, to geology, to astronomy, their story stretches right back through the history of the universe, to the moment of creation itself.

Astronomers now have extremely convincing evidence that everything in the universe can trace its ultimate origin back to the violence of the Big Bang, 13.8 billion years ago, when space, time and everything in it sprang into existence. Today, when we look out into the depths of space, we see that distant galaxies are receding away from us as space itself expands, carrying them with it as it goes. Rewinding this expansion inevitably leads us to the conclusion that the galaxies must have been closer together in the past and, going back further still, that everything must have emerged from a single point of unimaginable density – a singularity.

Other studies show that the entire universe is also pervaded by a low-level 'hiss' of microwave radiation, corresponding to an ambient temperature of 2.725 degrees above absolute zero. Extrapolating back into the smaller, denser universe of the past, this Cosmic Microwave Background Radiation would have been hotter and more intense – and we now understand that it is the afterglow of the Big Bang itself, spread out and cooled down from its initial incandescent levels by billions of years of expansion. These observations have been repeated and verified many times and it is extremely hard to think of an explanation for them that doesn't involve a hot, dense beginning for the universe. The Big Bang is one of the best-supported theories in physics.

In its first few instants the universe was infinitesimally small and consisted of a blaze of raw energy, but as it expanded and cooled from this unimaginably hot and dense initial state the most fundamental particles of matter – quarks, gluons, electrons and others – began to condense. At first, these particles bounced and ricocheted off one another but, as the temperature continued to fall, their collisions became less and less violent, eventually enabling some of them to stick and join together, forming larger, more massive particles. Just below 2 trillion degrees Celsius, quarks began to clump into groups of three to form protons and neutrons – the building blocks of atoms. A single proton is the central part – the nucleus – of a hydrogen atom, so it could be said that this marks the point in cosmic history when the first chemical elements began to form. The universe was one second old and it had grown to around a thousand times the size of our Solar System.

Hydrogen is the simplest and lightest element of all and on its own it isn't much use for making planets or people. To make other, heavier elements, individual protons need to join together with neutrons and with other protons to form larger and more complex nuclei. When the universe was a few seconds old, its temperature had dropped to just over 100 billion degrees. Now, at last, the protons and neutrons themselves could combine and the production of heavier atomic nuclei really got going. First of all, protons and neutrons joined to form pairs – the nuclei of a heavier type of hydrogen atom called deuterium. Deuterium was the crucial step in the formation of even heavier elements and it is still a very useful element today: water made with deuterium rather than ordinary 'light' hydrogen atoms is known as 'heavy water' and is used in medical tests as well as some types of nuclear reactor. From deuterium, which has one proton and one neutron, the next element to be made was helium, which has two protons and two neutrons. Small amounts of lithium (three protons and four neutrons) and beryllium (four protons and five neutrons) also began to form.

At some point during these first few minutes, another type of particle was also created. This was 'dark matter', and it was made in large quantities – five times as much of it as the 'ordinary matter' of electrons, quarks, protons and neutrons. Despite this abundance, we still don't know for sure exactly what sort of particle dark matter consists of, or exactly when it came into existence – but it seems likely to have been around the same time as the other particles and, as one of the most significant components of the universe by mass, its gravity would go on to play a defining role in helping ordinary matter clump together into galaxies, stars and planets.

But this was all still a long way in the future. Only 20 minutes after the Big Bang – just as things were really getting interesting – the temperature of the universe fell to below 1 billion degrees Celsius and the atomic production line ground to a halt. Collisions were now too rare and too gentle for any new atomic nuclei to be formed and the atomic composition of the cosmos was fixed, for the time being at least. The universe now consisted of a seething cloud of atomic nuclei, electrons and other particles of matter all mixed together in a fog of electromagnetic radiation consisting of photons – the particles that make up light. This soup of matter and energy was dense enough that individual photons regularly collided with particles of matter and were bounced in all directions – in other words the light could not travel unimpeded for large distances, so the universe was therefore opaque.

About 378,000 years later the universe had grown to around one thousandth of its current size and its temperature fell below another critical threshold. At just under 3,000 degrees Celsius, each positively charged atomic nucleus was able to grab and retain a complementary cloud of negatively charged electrons. Together, nuclei and electrons formed complete atoms, and the first chemical elements had arrived. This transition from sub-atomic particles to fully fledged atoms also had another profound effect. As the positively charged nuclei cloaked themselves with negatively charged electrons, the atoms that they formed were electrically neutral. Neutral atoms interact much more weakly with electromagnetic radiation than charged particles do: suddenly the photons that had been furiously bouncing between the particles of matter for 378,000 years were free and, instead of an opaque fog of light and matter, the universe was transparent for the first time. These photons are still all around us today – they form the Cosmic Microwave Background (CMB) radiation, which provides one of the strongest pieces of evidence for the Big Bang itself (*see* White Dielectric Material, page 39). The expansion of the universe has stretched and cooled them and they now lie mostly in the microwave part of the spectrum – at wavelengths far too long for our eyes to see but still detectable by radio telescopes. When a photon from the CMB is picked up by one of our telescopes, this is usually the first time it has interacted with particles of matter since the universe became transparent all those billions of years ago: its photons have travelled further to reach us than any other radiation we can detect and so the CMB is our direct link to the time when the first atoms formed.

Atoms are the building blocks of ordinary matter – the familiar solids, liquids and gases that make up stars, planets and living things – but this brand-new atomic universe was different in one very important way from the one that we inhabit today. That first 20-minute burst of nucleosynthesis produced only the very lightest and simplest elements in the periodic table: atoms of hydrogen and helium and a trace of lithium atoms. Tiny amounts of beryllium had also been formed in these first minutes, but these atoms were radioactive and within a few weeks they had decayed, producing more lithium. This was not an auspicious start, at least not from the point of view of carbon-based lifeforms like us, living on a wet, rocky planet. The complex chemical makeup of today's universe is an essential part of what makes life here on Earth possible. Without silicon, there would be no silicate rocks to make planets, without oxygen there would be no water to sustain life and without carbon there would be no organic chemistry on which to base it. We need iron for our blood and calcium for our bones, and a myriad of other elements to build the complex chemical machinery that makes up a living organism.

Today, we're familiar with 98 naturally occurring chemical elements, ranging from hydrogen, the lightest and simplest of all, with one proton and one electron, through to californium, which has 98 protons, 98 electrons and (usually) 153 neutrons. About half of these elements, particularly the heavier ones, are radioactive, which means that the nuclei of their atoms – where the protons and neutrons are found – are unstable and will eventually disintegrate, to produce nuclei of several lighter 'daughter' elements. For many elements the timescale over which this happens can be millions or even billions of years, although for others it is much shorter – and these short-lived elements are correspondingly less common. In addition to the 98 elements found in nature, around another 20 very massive but highly unstable elements are also known to exist. Their atoms break apart so quickly that they have only been fleetingly observed here on Earth – and only when scientists have managed to synthesize them in tiny quantities in the laboratory. It is likely that other, still more massive, elements may also be allowed by the laws of physics.

The atoms that coalesced from the soup of electrons and nuclei when the universe was 378,000 years old formed a bland expanse, consisting of 1 helium atom for every 12 hydrogen atoms, with a very light sprinkling of lithium atoms mixed in. Because each helium atom has four times the mass of a hydrogen atom, measured by mass this expanse was made up of 75% hydrogen, 25% helium and a ten-millionth of a per cent of lithium. Today the atomic inventory of the universe is not radically different: hydrogen makes up 74% and helium 24% of all atoms by mass, but, crucially, 2% of the atomic matter in the universe now consists of the heavier elements that make up the rest of the periodic table. By far the most abundant of these are oxygen and carbon – essential for water and life – followed by neon, iron, nitrogen, silicon, magnesium and sulphur.

Despite their relative scarcity, the physical and chemical properties of these heavy elements allow them to interact with each other and to clump together, becoming more concentrated. The Sun and the gas giant planets like Jupiter and Saturn have a similar elemental abundance to that of the universe as a whole, being dominated by hydrogen and helium, but rocky planets like the Earth are super-concentrations of the rarer, heavy elements. By mass, the Earth is composed of 32% iron, 30% oxygen, 15% silicon, 14% magnesium, 3% sulphur, 2% nickel, 1.5% calcium and 1.4% aluminium, while the remaining 1% consists of trace amounts of the other elements.

Meanwhile, again by mass, the oceans consist of 86% oxygen and 11% hydrogen (between them, making water), 2% chlorine and 1% sodium (salt), with traces of magnesium, sulphur, calcium and potassium. The atmosphere is 78% nitrogen and 21% oxygen, with the rest composed of argon, carbon (in the form of

carbon dioxide) and, depending on how humid the air is, varying amounts of hydrogen (in the form of water vapour). Even
closer to home, the mass of a human body is 65% oxygen (mostly
bound up in water molecules), 18% carbon, 10% hydrogen
(bound up in water and organic hydrocarbon molecules), 3%
nitrogen, 1.5% calcium 1.2% phosphorus and a huge range of
essential trace elements which, while only present in small
quantities, all play a vital role in our metabolism.

Clearly, neither our local environment nor our own bodies could
have formed from the mix of elements created by the Big Bang.
Where did the rest of the elements come from? How did the
universe go from a sterile expanse of hydrogen and helium to
the cocktail of elements and molecules that we find today? The
answer to these questions is plainly visible, shining above our
heads in the form of the Sun and the other stars.

Today, each individual star, deep inside its heart, recreates those
hot, dense conditions of the early universe, sustaining them not
just for minutes, but for millions or even billions of years – and
thus allowing the nuclear reactions that build up heavier
elements to fire up once again. The stars are the factories in
which all of the elements heavier than lithium have been forged,
and our planet and everything on it – including us – is composed
largely of these heavy elements, which were built up from the
primordial hydrogen and helium inside generation after generation of stars.

It is gravity that enables the stars to generate such extreme
conditions in their cores. A typical star like the Sun contains

around 2,000,000,000,000,000,000,000,000,000,000 (2 million trillion trillion) kilograms of matter, mostly in the form of hydrogen and helium, and the crushing weight of all this material bears down on the centre of the star, squeezing it tightly and producing the intense pressures and temperatures required for nucleosynthesis to resume. Here, energetic collisions between the basic atomic building blocks of hydrogen and helium nuclei lead by various combinatory routes to the production of carbon, nitrogen, oxygen and other, heavier elements – a process known as 'nuclear fusion' since it involves the fusing together of light nuclei to create heavier ones. The more massive the star, the higher the central pressures and temperatures and the more energetic the collisions between particles: massive stars are therefore able to produce a wider range of elements.

Each time a heavier atomic nucleus is created from the collision of two lighter ones, a tiny amount of their combined mass is converted into energy in accordance with Einstein's famous equation $E=mc^2$, where E is the energy released, m is the amount of mass lost and c is the speed of light. Although the mass lost in each fusion event is very small, the speed of light is very large – and a large number squared is a very large number indeed. The energy is released mostly in the form of electromagnetic radiation – as a high-energy photon in the gamma ray part of the spectrum. In the dense conditions at the centre of the star, these photons ricochet off nearby particles of matter, transferring some of their energy in each collision and raising the temperature and pressure of the core still further as they fight their way out of the star.

It is this additional temperature and pressure that keeps the star stable as the outward pressure of the superheated core supports the star's upper layers against the immense gravity that is trying to force them inwards. Every star in the sky, including the Sun, is a delicate balancing act between these inward and outward forces: we can think of them either as a mass of collapsing gas held up by its own internally generated pressure, or as a nuclear explosion prevented from expanding outwards by the weight of its own outer layers. Far from the serene and unchanging orbs we might fondly imagine them to be, each star is a straining knot of contradictory forces, caught in an ongoing struggle between the urge to collapse and the urge to explode. Meanwhile, the photons make their laborious way outwards, through the star's upper layers, losing energy in each encounter with a particle of matter, attenuating from gamma rays to X-rays and eventually flooding out into space as photons of visible light, and ultraviolet and infrared radiation – the star shines. Like the atoms in our bodies, starlight has its origins in the nuclear furnace deep within each star – it is a by-product of the elemental production line that also created us.

No one knows exactly when the first stars formed from the primordial clouds of hydrogen and helium, but it was probably around 500 million years after the combination of electrons and atomic nuclei into atoms. Before these early stars began to shine the universe had been dark, with very little new electromagnetic radiation being produced, ever since the photons of the CMB had been released on their long journeys. Time was required to allow gravity to do its work: the primordial matter was spread quite evenly through space, but slight variations in density acted as gravitational seeds, gradually pulling more matter towards them, and the process fed on itself, forming denser and denser accumulations. It was here that dark matter, five times as abundant as all the hydrogen and helium atoms combined, played a vital role, dominating the clumping process and pulling the hydrogen and helium atoms along with it. Eventually, large clouds of gas had gathered – the embryos of today's galaxies – and within them were concentrations of sufficiently high density to collapse still further, forming individual stars. The universe began to shine once more.

This very first stellar generation would not have been quite the same as the stars we see around us today, however. Formed from a very pure mix of hydrogen and helium, with just a trace of lithium, the earliest stars would have balanced the gravity outside and the pressure inside in a slightly different way from that of their modern counterparts. Scientists suspect that they would have been real giants, containing several hundred times as much gas as a star like the Sun and blazing millions of times more brightly. But stars would never again be quite this big or this bright. As soon as elements heavier than hydrogen and helium joined the mix in significant quantities, the stars' internal balance mechanism was changed forever: today, a star this massive would not be able to hold itself together against the outward pressure of its own radiation and it would be blown apart before it could properly form. The current upper limit on the mass of an individual star seems to be around a hundred times the mass of the Sun, but perhaps this is just as well. If such primitive stellar giants were still around today, their harsh radiation and stupendously violent death throes would make conditions extremely uncomfortable for life on any nearby planets.

It is an inevitable fact of stellar physics that very massive stars live fast and die young, racing through their supplies of nuclear fuel and blazing radiation into space at much higher rates than an average star like the Sun due to the immense pressures and temperatures generated in their cores. This would have been particularly true of the earliest stellar giants – the equations that govern the structure and lifespan of stars suggest that they could only have lasted for a few million years before reaching the end of the line. But even these relatively brief lifespans would have been enough to generate significant quantities of heavy elements, and within a few million years the elemental mix of the cosmos had been changed forever, with every subsequent generation of stars since then further enriching the brew. By the time our Sun began to form, 9.3 billion years after the universe itself came into being, there were sufficient amounts of heavy elements in its vicinity to give rise to a planet with an iron core encased in silicate rocks, with oceans of water, and an atmosphere of nitrogen and oxygen – as well as the complex, carbon-based molecules necessary for life.

But once all these essential elements had been produced, how did they get out of the stars where they were made and into the wider universe? Unless the products of nuclear fusion can be liberated and spread far and wide, subsequent generations of stars – and planets – would never benefit from their creation. The solution, so often the case in nature, is that the enrichment of the next generation is achieved by the death of the previous one.

The lifespan of a star is determined at birth and depends almost entirely on its initial mass. This dictates the strength of the star's gravity, and thus the temperature and pressure in its core and the rate at which nuclear fusion reactions proceed. Even a small star is a colossal object, and stellar cores contain a great deal of hydrogen to act as nuclear fuel, but nothing lasts forever and eventually all of the hydrogen in the core will be fused into helium and this initial supply will run out. What happens next once again depends on the mass of the star. As its nuclear furnace stalls, less radiation is generated in the core and the star's delicate balance will tip in favour of the inward force of gravity. Crushed by the weight of its outer layers, the core – now a giant ball of helium 'ash' – will begin to contract. But this contraction will raise the core's temperature and pressure still further until, if the star is massive enough, it reaches the point at which helium begins to fuse into carbon.

With renewed nuclear reactions releasing a fresh supply of energy, the star's balance is restored and the core stabilizes once more – at least until this new supply of fuel is exhausted in its turn. The process repeats, this time with carbon and helium fusing together to produce neon, then neon and helium combining to form oxygen and so on through the first twenty elements of the periodic table, many of them essential for life. In this way, a star's life consists of a series of episodes in which the core shrinks, raising the temperature and pressure to the point at which it can begin a new cycle of fusion using the products of the old as fuel. As the star matures, its core takes on a complex structure of nested shells, like the layers of an onion, with successively deeper shells fusing heavier and heavier elements. The process continues until the gravitational squeezing of the core is no longer sufficient to trigger the next fusion cycle – a fundamental limit set by the star's total mass.

The Sun, a star of modest mass, will falter at a relatively early stage in this sequence, having fused helium into carbon and oxygen but progressing no further. However, it will do this at a sedate pace and the process will take around 10 billion years in total. At 4.5 billion years old the Sun is currently just under halfway through its life as a fusion factory; it still has an exciting career ahead of it – one that we will follow in more detail in a later chapter, since it will have profound consequences for the future of our planet.

Stars of greater mass than the Sun will continue through further cycles of fusion – and at a more rapid pace than the Sun – but all will ultimately reach their pre-ordained endpoints. When fusion finally grinds to a halt the star's internal balancing act, carefully maintained for so long, will also come to an end, as gravity and radiation pull its inner and outer layers in opposite directions and the star is gradually teased apart.

The inner core will succumb to gravity, contracting down into a sphere not much larger than the Earth, but incredibly dense and composed of the debris of successive cycles of nuclear fusion compressed into a strange state known as degenerate matter. Unable to generate a continuing supply of energy via nuclear fusion the core is now supported by the pressure of electrons, whose quantum mechanical properties prevent them from being squeezed any closer together. This shrunken remnant is known as a white dwarf star and, although its residual heat will keep it shining for billions of years to come, it will gradually cool and fade into a cold, dead stellar cinder.

As the core shrinks towards its compact end-state, the star's outer layers, also laden with fusion products, will suffer the opposite fate as they are ejected gently into space over thousands of years to form a vast, glowing cloud several light years across, known as a 'planetary nebula'. As the nebula continues to expand, it will eventually merge with the interstellar medium – the mix of gas and dust that drifts between the stars – and this is one of the main ways in which the wider universe becomes enriched with heavy elements. At some point in the future, pockets of this interstellar gas will clump together once more and gravity will shape them into another generation of stars to begin the process anew. In this way, the elemental mix of the Solar System is the product of many previous cycles of star birth and star death, and when our Sun finally dies it too will bestow a gift of newly forged heavy elements to future stellar generations.

COSMIC DATING AGENCY

On a dark, moonless night, far from city lights and other sources of light pollution, the sky is dominated by the majestic band of the Milky Way – a ribbon of pale, hazy light that seems to circle the entire sky. In the seventeenth century, Galileo Galilei used a telescope to demonstrate that the Milky Way's misty glow was in fact the combined effect of millions of stars, too faint and too numerous for the naked eye to resolve them individually. We now understand that the Milky Way is a spiral galaxy, a flat disc over 100,000 light years in diameter and containing perhaps 200 billion stars. Our Solar System lies within the thickness of the disc and so we see the majority of the other stars in the galaxy arrayed around the sky in a vast circular band.

But even with the naked eye an inspection of the Milky Way quickly shows that its light is not evenly distributed. Indeed, some parts of the band appear to be incised with dark 'holes', almost as if the disc of the galaxy contains voids in which there are hardly any stars at all. In fact, this is a false impression: if we observe the sky with telescopes that are sensitive to infrared radiation rather than visible light, we find that these apparent voids are just as full of stars as the surrounding regions. The reason that they appear dark is not because they are empty of stars, but because they are full of something else: great clouds of space dust that absorbs and blocks all traces of visible starlight but which allows infra-red radiation to shine straight through.

Once considered an annoying obstruction by astronomers who were eager to see the distant reaches of the Milky Way, we now know that this dark cosmic dust is an integral component of our own and other galaxies – and it is just as interesting as the stars and clouds of glowing gas, with their conspicuous displays of light. Known as 'giant molecular clouds', these dark regions perform a vital role in the anatomy of the Milky Way. On an atomic scale, the dust acts as a galactic chemistry lab in which a vast array of complex molecules are cooked up from the raw ingredients of the periodic table. But its influence is not limited to the world of atoms and molecules: as we have come to realize, it also has a part to play on the scale of the galaxy itself, nurturing the formation of new generations of stars and planets.

As we've seen, the stars are atom factories, building up heavier and heavier elements by fusing lighter ones together, but there is more to creating planets – and to sustaining life – than the 98 naturally occurring elements of the periodic table. Most of the complexity of geology, chemistry and biology is ultimately based on the combination of individual atoms to form molecules: silicon and oxygen to make silicate rocks, hydrogen and oxygen to make water, carbon and hydrogen to make the thousands of organic compounds that form the building blocks of living things. But in order for atoms to link together in this way, they first of all need to meet up with each other – and in the vast expanses of space, this can be enormously challenging.

Compared with the majority of the universe, here on Earth we are blessed with a remarkable abundance of matter: our local environment is awash with atoms and molecules, gathered together as solids, liquids and gases and in concentrations many billions of times greater than the cosmic norm. We think of the air that surrounds us as the very definition of lightness and intangibility, and yet each cubic centimetre of it contains around 30 million trillion atoms and molecules of nitrogen, oxygen, argon, carbon dioxide, water and various other substances. By contrast, in a typical cloud of interstellar gas, drifting between the stars, the same cubic centimetre would contain one solitary atom of hydrogen. In the depths of space such an atom could wait for a very long time before encountering a partner with which to form a molecule.

Of course, there are other places in the universe besides Earth where the density of matter is much higher than average. Deep inside the stars, hydrogen, helium and all of the newly generated heavy elements are present in close proximity, where they are constantly colliding with each other; but at temperatures of millions of degrees these collisions are far too violent to allow interesting molecules to form. Only in the outermost layers of the stars, where temperatures can be as low as 3,000 degrees Celsius, can atoms come together gently enough in order to stick together – red dwarf and red giant stars are some of the universe's most unusual chemistry laboratories and since the 1990s scientists have observed evidence for a variety of molecules in their cool stellar atmospheres. Surprisingly, even the Sun, with a surface temperature of around 6,000 degrees Celsius, plays host to a handful of robust molecules, including magnesium and calcium monohydrides and titanium monoxide.

But chemistry in the outer layers of the stars is limited to the most heat-resistant molecules. By contrast, in order to remain stable, most molecules require the cooler conditions that can only exist far from the harsh blast of stellar radiation, and here we return to our original problem: space is cold, but it is also very empty and under these rarefied conditions chemistry is a very slow process indeed. A helping hand is required in order to get the molecular production line moving, and this is where cosmic dust plays a vital role.

Cosmic dust is somewhat different from the powdery deposits that accumulate in our homes here on Earth. Several types have been identified, formed under a variety of different astronomical conditions but all are very small, with typical grain sizes just a few millionths of a metre across – comparable in size to smoke particles. Many grains seem to have their origins in the harsh conditions of stellar atmospheres – and even in the incandescent debris of supernova explosions – where hardy substances such as silicon carbide (otherwise known as the mineral carborundum, used in ceramics, brake pads and bulletproof vests), aluminium oxide (the basis of sapphires and rubies), silicates, graphite and diamond are able to condense into tiny solid particles. As they are blown out from the star itself, riding into space on the stellar wind, they begin to acquire a veneer of less heat-tolerant materials, finally becoming encased in an icy layer of frozen water and other volatile substances. Eventually, the dust grains become mixed with the gas that drifts between the stars, and these vast regions of intermingled gas and dust are the dark patches that appear to mar the starry expanse of the Milky Way. Deep inside such dusty clouds the scene is set for the formation of rocky planets like the Earth – and for the origins of life itself.

Cosmic grains may be small by our standards, but compared with the tiny scale of an atom they are like mountains floating in the vacuum of space. A lonely atom has a far higher chance of bumping into a bulky dust grain than it does of encountering another atom. But, once it has hitched a ride on the icy surface, suddenly all of the dust's other atomic passengers are within easy reach – and available for any number of chemical liaisons. For this reason, dust grains have been dubbed 'cosmic dating agencies' where atoms can meet and combine into a wide range of molecules.

The chemical connections facilitated by the dust are responsible for a growing list of molecules, whose signatures have been detected far out in space, ranging from simple compounds such as carbon monoxide, sodium chloride, hydrogen sulphide and nitrous oxide, to surprisingly complex molecules like acetic acid, urea and the strange, spherical carbon molecule known as buckminsterfullerene. Altogether, more than 200 different molecules are already known to be widespread in the clouds of gas and dust that thread the galaxy, and a discipline known as 'astro-chemistry' has sprung up in order to study them. Among these plentiful extraterrestrial substances are water, carbon dioxide and ethyl alcohol – so it is reassuring to know that the basic ingredients of a vodka and tonic are common throughout the universe. Even more reassuringly, many of the organic molecules thought to be prerequisites for the chemistry of life are also common in space. It's important to stress that, in chemistry, the word "organic" simply means a member the huge family of molecules containing carbon atoms. All living things are based on organic molecules but not all organic molecules are involved in living things. However, of the organic molecules known to exist in space, many of them are indeed thought to be prerequisites for the chemistry of life, including polycyclic aromatic hydrocarbons and the substance pyrimidine, from which several building blocks of the DNA molecule can be derived.

As well as playing matchmaker for the formation of these life-giving molecules, there is a further reason for us to be grateful to cosmic dust. Molecules floating in space are all very well, but in order for them to give rise to planets and thence to life they need to be brought together once more in the warm conditions close to a star. Gravity is the main force by which this feat is accomplished but, once again, the dust provides a helping hand.

Slight variations in the density of interstellar gas clouds are the seeds that lead to star formation: more gas is attracted to the densest regions of the clouds, making them denser still and initiating a process of gravitational collapse. But the laws of physics are strict. As a gas is squeezed into a smaller volume it also warms up – an effect that we can observe here on Earth when air is compressed inside a piston. As the interstellar clouds contract under their own gravity, they too inevitably become warmer – and eventually the rising temperature and pressure of the gas can balance out the force of gravity, halting the collapse and causing the star formation process to stall. Some means is required for the gas to get rid of this excess heat before it can condense further, and this is where cosmic dust comes into play. Embedded within the gas the dust grains act as trillions of tiny radiators, venting the excess energy into space in the form of infrared radiation. With this in-built cooling mechanism in place, the barrier to further collapse is removed and the path to forming new stars and their attendant planets is clear once more. The birth of new stars is a process driven by gravity, but without the dust to act as midwife it is likely that the Milky Way's population of stars and planets would be much smaller.

Here on Earth we think of dust as a trivial annoyance – untidy perhaps, but hardly worth our serious attention. However, the grains of dust that float throughout the Milky Way are quite the opposite: without them the Sun, the Earth and the molecular building blocks of life would probably not exist.

The process of fusion in ordinary stars like the Sun accounts for the existence and wide distribution of the first twenty elements, but what about the other, heavier elements of the periodic table? A clue comes from a faint patch of light in the constellation of

Taurus. Invisible to the naked eye, the Crab Nebula was first observed telescopically by the English doctor and astronomer John Bevis in 1731 and was discovered again, completely independently, by the great French astronomer Charles Messier in 1758. Messier was actually looking out for the return of Halley's Comet, which had been predicted to revisit the inner Solar System that year – arriving from the direction of Taurus. Through a telescope the cloudy, fuzzy patch of light looked superficially like a comet and it was only by observing it for several nights that Messier could be sure that it remained fixed in position rather than moving against the background stars in genuine cometary fashion. To avoid any further false alarms Messier resolved to make a catalogue of these fixed *nebulae* (Latin for 'clouds') so that astronomers could ignore them and get on with the highly competitive (and far more glamorous) business of discovering new comets. The cloudy patch of light in Taurus duly became Messier 1, or M1 for short – the first object on Messier's list.

The Messier Catalogue is still widely used today, but astronomers now regard its contents as fascinating objects in their own right rather than distractions to be avoided. Even so, it was some time before the true nature and significance of M1 was realized. The nebula didn't even acquire a more evocative name until 1844 when the aristocratic astronomer William Parsons, Earl of Rosse, sketched it through one of the powerful telescopes he'd installed at his observatory at Birr Castle in Ireland. Rosse felt that his drawing displayed a distinctly crustacean appearance and coined the name 'Crab Nebula' but, after observing it through a more powerful telescope four years later, even he was hard pressed to see any convincing resemblance to a crab. Certainly, rather than a claw-wielding sea creature, modern photographs show a distorted, bubble-like structure formed from delicate filaments of glowing gas, but Rosse's original name has stuck.

Whatever name we decide to give the nebula, one thing is for sure: photographs of it taken several decades apart clearly show that it is expanding, with gas moving outwards in all directions at the colossal rate of 1,500 kilometres per second. By tracing this motion backwards we can estimate the date of the explosive event that flung all of this material out into space and we find that it must have taken place around a thousand years ago. Here, we are indebted to the meticulous record keeping of medieval Chinese astronomers who in 1054 noted the appearance of a dazzling new star in exactly the same part of the sky. Almost as bright as the Moon, this temporary 'guest star' was visible to the naked eye for up to two years before finally fading away, and at its peak could even by seen in the daytime sky. The Chinese astronomers had observed a supernova – the explosion of a massive star – and the Crab Nebula is its scattered debris, still spreading outwards almost a millennium later. Together, the Chinese guest star and its expanding remnant are the missing link that explains the origin of all the chemical elements heavier than iron.

Nuclear fusion inside the stars is self-sustaining because each time two light elements are fused together to produce a heavier one the reaction releases energy in the form of a photon. This energy release maintains the star's internal balance between the inward force of gravity and the outward pressure of the hot gas

– and the photons themselves enable the stars to shine steadily for millions of years. Energy can be released in this way during the formation of all the elements in the periodic table up to and including iron, the 26th element, which has one of the most stable atomic nuclei of all. This stability makes iron a pivotal element in the physics of very massive stars, but it is also their downfall. Iron can be fused into heavier elements but these reactions soak up energy rather than releasing it – and this reversal catastrophically undermines the delicate equilibrium between pressure and gravity that has kept the star stable throughout its life. The consequences are devastating for the star – but without them we humans would not be here at all.

Most stars are incapable of producing iron: they are simply not massive enough to generate the necessary crushing pressures and 2.5-billion-degree temperatures within their cores. But a small fraction of stars – those with masses of more than 8 times that of the Sun – are able to achieve this feat, allowing them to continue fusing their way through the elements of the periodic table. For them iron is the inevitable end product of nuclear fusion as their cores progressively switch their main fuel source from hydrogen to helium, carbon, neon, oxygen and finally silicon, generating other elements along the way as side products. Each stage yields diminishing returns as the amount of energy released per reaction dwindles, and the star's internal balancing act gets harder and harder to maintain. When the star begins to fuse silicon into iron and nickel, the tipping point is in sight. A core of iron and nickel 'ash' begins to build up at the heart of the star, growing in mass but generating no new energy. This iron is at much higher temperatures and pressures than the iron core of the Earth: instead of a molten liquid or a crystalline solid, the stellar iron exists as a plasma – an extreme form of matter in which the electrons are stripped from their atoms and exist in a kind of 'soup' of particles along with the remaining atomic nuclei. Like the material in a white dwarf star, this core is temporarily supported against gravity by the degeneracy pressure of its own electrons but this cannot persist indefinitely.

As more and more iron and nickel are created, the mass of the core increases until it exceeds 1.4 times that of the Sun. This is the critical point – the core of the star implodes, collapsing inwards at more than a fifth of the speed of light. Even electrons can no longer resist the overwhelming force of gravity: they are crushed into the protons that make up the iron nuclei, forming neutrons and releasing a flood of tiny particles called neutrinos. Within a few milliseconds, the temperature within the collapsing core reaches 100 billion degrees Celsius and its particles are squeezed together at densities similar to those in an atomic nucleus. The core has become a neutron star – an object containing more matter than the Sun but compressed into a super-dense ball just a few kilometres across (*see* On Stronger Tides, page 190).

If the mass of the core exceeds five times the mass of the Sun then even the degeneracy pressure of neutrons will not be enough to support it against the inwards force of its own gravity. Instead of halting as a neutron star a few kilometres across, the core will continue its relentless contraction, squeezing itself down to a point of zero size and infinite density – a state known in physics as a 'singularity'. With the mass of several Suns condensed into an infinitesimal volume this ultra-compact object exerts an intense gravitational field on the space around it. Within a few kilometres of the singularity gravity is so strong that even light – travelling at the ultimate cosmic speed limit – isn't moving fast enough to escape from its vicinity: the stellar core has become a black hole. Only stars with an initial mass of greater than about ten times the mass of the Sun will end up as black holes – such stars are rare, but already our telescopes have detected more than a dozen black hole candidates in the Milky Way alone.

The collapse of a massive stellar core into a a neutron star or a black hole is a sudden event and, in contrast to the leisurely contraction of an ordinary star into a white dwarf, a massive core will shrink from tens of thousands of kilometres in diameter to a tiny size in just a few seconds. As the core stabilizes in its new compact form, a huge burst of energy is released, and a violent shockwave propagates outwards towards the surface of the star. At the same time, the intense burst of neutrinos generated by the crushed material also races outwards, helping to drive the shockwave with even greater force. Within seconds, the implosion of the core leads to the explosion of the rest of the star, ripping the outer layers apart and propelling the debris into space in a blaze of electromagnetic radiation. For a period of several weeks to months the expanding debris shines brightly as a supernova. (*See also* Exploding Stars, box, page 319.)

It is in the chaos of the explosion that new elements heavier than iron are forged. These reactions require energy, but now energy is in abundant supply and as the shockwave of the explosion passes through the outer layers of the star the gas is compressed and heated to the point at which nuclear fusion can resume. Within seconds a vast new array of heavy elements is created, from arsenic, cobalt and copper to iodine, lead and zinc – and on up to the heaviest naturally occurring elements of all, such as uranium.

As the light of the explosion fades, this debris spreads out into space, becoming a nebula like the Crab and eventually merging with the interstellar medium, enriching it with all the elements of the periodic table.

It turns out that not all Sun-like stars are guaranteed an uneventful life once they have discarded their outer layers and settled down to retirement as a white dwarf. Unlike the Sun, around a third of all the stars in the Milky Way exist as binary pairs, in which two stars are gravitationally bound in orbit around each other. If a white dwarf in a binary system is close enough to its companion it can sometimes siphon off hydrogen from the second star's outer atmosphere, cloaking its own compressed and degenerate heart in a shell of fresh nuclear fuel. Once this hydrogen shell reaches a critical mass the whole lot will detonate, undergoing a catastrophic burst of nuclear fusion that blows the entire star apart. This is a rather different scenario from the collapse of a massive stellar core but the result is just as violent and for historical reasons such exploding stars are known as Type 1A supernovae (the more traditional core-collapse supernovae are labelled as Type 2). In fact, astronomers are beginning to suspect that many Type 1A supernovae might be the result of an even more violent scenario: the collision and merging of a binary pair consisting of two white dwarf stars.

Unlike the collapse of a very massive star, a Type 1A supernova leaves no compact remnant either in the form of a neutron star or a black hole. Instead the entire object is disrupted, with the original material of the white dwarf being blasted outwards and fused into heavier elements in the process – a second, and rather more comprehensive demise for the original Sun-like star. In a Type 2, core-collapse event, much of the iron produced by the massive star before it exploded will remain trapped inside the compact remnant for evermore. It's therefore likely that Type 1A explosions are the source of the majority of the iron that we find around us here on Earth.

Between them, both types of supernova have produced all of the elements heavier than iron, scattering them back out into space. Among them are atoms of the metals platinum, silver and gold and so every time we wear jewellery made from these precious materials we are adorning ourselves with the debris from some of the most violent events in the universe – atoms that share a direct kinship and origin with the exotic ultra-condensed matter that lurks within neutron stars and black holes. The same is true of the iron that gives our blood its vivid red colour: this vital living essence links us to the death throes of stars that went supernovae billions of years ago.

The debris of these explosions spreads out into the galaxy, gradually mixing and merging with other clouds of interstellar gas and dust. At some point during their random swirling and eddying some regions of the clouds will achieve sufficient density for gravity to take hold once more. Here too, other supernova explosions may play an important role: the shockwaves generated as their debris collides with nearby gas clouds may provide the extra boost required to compress the gas and trigger its collapse.

Slowly at first, but with increasing urgency, the clouds will begin to clump and condense, and the densest knots will ultimately become the seeds of a new generation of stars. We can see this process occurring today in active star-forming regions such as the Orion Nebula. Indeed, one of the most famous astronomical photographs of all, the Hubble Space Telescope's 'Pillars of Creation', shows this process in action: vast columns of gas and dust several light years tall are capped with dense clumps which act as nurseries in which new stars are being born.

Thanks to the lives and deaths of their stellar predecessors this new generation of stars will be richer in heavy elements and – crucially – so will the rotating discs of leftover dust and gas that swirl around them as they form. Within these 'circumstellar discs' solid grains of dust collide and stick together, forming pebble-size objects, then boulder and mountain-sized clumps – building up into asteroids and protoplanets. As they continue to grow, the strengthening gravity of the developing planets allows them to draw in the surrounding gas, cloaking themselves in the beginnings of an atmosphere.

Eventually, the core of each young star ignites its fusion fires for the first time, embarking on a fresh career of synthesizing heavy elements that will last for millions or even billions of years. As the new-born star begins to shine, the torrent of electromagnetic radiation drives off the remaining gas and dust around the star, shutting off the process of planetary growth. The stellar lifecycle has come full circle: out of the death of previous generations a new star, complete with orbiting planets, has arrived and, if conditions are right, on some of those planets the stage may be set for life to emerge. This is how our Sun and its solar system were formed 4.5 billion years ago, from atoms created in the hearts of long-dead stars.

Centuries of astronomical observation have taught us that the stars in the night sky are almost unimaginably remote. This is indeed the case, and yet astronomy has also revealed that, in one sense at least, these distant lights are as close to us as our own flesh and blood. The atoms that make up our bodies – the oxygen, carbon, nitrogen, calcium and all the other elements on which our biology depends – were forged inside the stars themselves. As the cosmologist Carl Sagan observed: 'We are made of starstuff.'

White Dielectric Material

Astronomy has a reputation for being a high-minded endeavour, carried out in remote observatories and far removed both literally and metaphorically from the mundane concerns of everyday life. As we've already seen, this is rarely true: the celestial objects and phenomena studied by astronomers can affect our lives in all sorts of surprising ways and the knowledge and understanding generated by astronomical observations can give us powerful insights into more immediate concerns back here on Earth.

But, as any astronomer can tell you, the work of observing the cosmos is not divorced from the mundane practicalities, the chance occurrences and even the indignities of everyday life. In fact, some of the most important turning points in the history of astronomy have occurred under circumstances that, with hindsight, must have seemed extremely inauspicious. This was certainly the case with one of the most profound and important discoveries in cosmology: the detection of the Cosmic Microwave Background – the afterglow of the Big Bang itself.

In 1964, Arno Penzias and Robert Woodrow Wilson, two young radio astronomers working for Bell Telephone Laboratories in the United States, were conducting experiments with a large, horn-shaped radio receiver at Holmdel, New Jersey. But probing the origins of the universe was not the main concern of Bell Labs. Instead the 6-metre-wide antenna had been designed to test an early satellite communication system called Project Echo, which – as its name suggests – involved transmitting microwave signals around the Earth by bouncing them off reflective balloons placed in orbit. This work had important practical and commercial implications. Today, it is hard to imagine a world without satellite communications but even in the 1960s – right at the beginning of the Space Age – it was already clear that satellites could revolutionize the speed, reach and reliability of telecommunications.

The two Echo satellites were enormous aluminium-coated spheres, each more than 30 metres in diameter. Designed to be highly reflective to signals in the microwave band – part of the radio spectrum – they were also extremely effective at reflecting sunlight and were clearly visible from the ground as bright, star-like objects moving across the sky. Both remained in orbit until the late 1960s, gradually losing altitude until they burned up in the Earth's atmosphere. However, unlike modern communication satellites, the Echo balloons were simply passive reflectors with no means of boosting the signals that were being bounced from them. To ensure that information was recovered back on Earth with maximum fidelity, it was essential that the receiving antenna down on the ground be made as sensitive as possible.

This is where Penzias and Wilson came in. With research backgrounds in radio astronomy, their expertise lay in detecting the faint natural radio emissions of distant objects in space. Like the reflected signals from the Project Echo satellites, natural sources of radio emission such as stars and galaxies can't be made to shine more brightly, so what you see is what you get. To maximize the quality of the signal you receive, your only remaining option is to increase the sensitivity of your receiving equipment.

Penzias and Wilson set to work on the Holmdel Horn Antenna. Having fitted it with their most sensitive receiver system to date, they were surprised to find that the signals they were trying to measure were contaminated by a source of interference that could not be explained. Methodically, they began to investigate and eliminate the possibilities. First to be rejected was the idea that the interference might simply be man-made radio noise from nearby New York City. The microwave noise was present whichever direction the antenna was pointing in the sky and at all times of the day and night, which ruled out a terrestrial origin.

Next, they decided to check whether the receiver itself might be the source of the interference: warm objects naturally emit a glow of microwaves, and the microwave noise being picked up by the receiver was equivalent to an object with a temperature of around 3 degrees above absolute zero. Lowering the temperature of the receiver should have reduced any microwave emission it was producing but, even with additional cooling in place, the mysterious microwave noise remained unaffected.

It was during their thorough investigation of the telescope that Penzias and Wilson discovered that they were not the only ones making use of it. A pair of pigeons had taken up residence inside the horn of the antenna itself, building a nest and leaving copious deposits of what the astronomers politely described as 'a white dielectric material'. A dielectric is a substance with specific electrical properties between those of a conductor and an insulator – and therefore capable of affecting forms of electromagnetic radiation such as microwaves. Pigeon droppings, rich in various salts, definitely fit this bill, and so it seemed that the astronomers had at last discovered the source of the mysterious microwave interference. The birds had to go and, given the famous homing abilities of pigeons, relocating them wasn't really a viable option: they were caught and humanely despatched.

The grisly deed accomplished, droppings were scraped from the antenna and observations recommenced. But the microwave interference had not gone away. Something, somewhere was glowing at a temperature of 3 degrees above absolute zero and if it wasn't down here on Earth then it must, after all, be up there in space – and in every direction too.

By coincidence, cosmologists had already been looking for just such a sky-wide glow for around 20 years. A debate had been raging between proponents of the Big Bang theory, which claimed that the universe had exploded from an unimaginably hot dense state billions of years ago, and the rival Steady State theory, which argued that the universe had always existed in much the same form as we see it today. In their favour, the Big Bang camp were able to cite the discovery by Edwin Hubble and others in the 1920s that distant galaxies appear to be receding from us – exactly as would be expected in a universe that was expanding from an initial explosion. But the Steady State model could be modified to include a continual stretching and expanding of space so the Big Bang cosmologists needed additional evidence in order to win the argument. Penzias and Wilson's microwave glow was exactly what they were hoping to see.

In the Big Bang scenario, the early universe would have been a hot, dense fog composed of electromagnetic radiation and particles of matter, all colliding and interacting with each other. But as the universe expanded, this matter and radiation would become more and more spread out, eventually – after about 378,000 years – reaching the point at which the radiation could pass freely through space without constantly having its path diverted by a collision: in other words the universe would become transparent for the first time. The radiation that would have been set free by this transition should still be shining throughout the universe today – filling the sky – but, because space has continued to expand over billions of years, it would have spread out and attenuated, cooling from its initial searing temperature of around 3,000 degrees to a chilly 3 degrees above absolute zero.

Penzias and Wilson's announcement of their mysterious microwave glow was met with great excitement in the world of cosmology. Here, at last, was the long sought-after Cosmic Microwave Background radiation, or CMB for short. The Steady State model had no really plausible explanation to offer for the CMB and so its detection was a major step in the establishment of the Big Bang as the leading model for the origin of the universe. Even today, with many more lines of evidence also lending weight to the Big Bang model, the existence of the

Cosmic Microwave Background remains one of the most persuasive arguments that our cosmos did indeed begin in a titanic explosion 13.8 billion years ago. This radiation is the afterglow of the superheated birth of the universe and it is still all around us today.

In 1978, Arno Penzias and Robert Wilson received the Nobel Prize for Physics in recognition of this milestone in our under-standing of the ultimate origins of everything – a milestone that arose from the unlikely starting point of a telecommunications project, and whose signature had initially been mistaken for the effects of pigeon droppings.

THE BIG BANG ON TV

You don't need a radio telescope full of pigeon droppings to see the Cosmic Microwave Background in all its glory. Although the bulk of this fossil radiation exists in the form of microwaves, there is in fact a range of wavelengths, and a small fraction of it comes in the form of ultra high frequency (UHF) radio waves of the type that we humans use for our analogue TV transmissions. Just as the CMB was first detected as static interference in the microwave signals picked up by the Holmdel Horn Antenna, so a small percentage of the static 'snow' on the screen of an un-tuned analogue TV is also due to this ancient background radiation.

It's astonishing to think that these microwave and UHF photons have been travelling uninterrupted for almost 13.8 billion years by the time they encounter our TV aerials and are converted into electrical signals to light up our screens. They have each endured what must be the longest and loneliest journey in the universe – but we needn't feel too sorry for them. Einstein's Special and General Theories of Relativity tell us that the faster an object moves through space the less it will experience the passage of time. This odd-sounding claim has been demonstrated again and again by numerous experiments: clocks sent rushing around the Earth at orbital speeds on board the International Space Station register less time passing than identical clocks down on the ground, while unstable subatomic particles projected at very high speeds inside particle accelerators are observed to last much longer before decaying than they would if stationary in the lab. At the kind of speeds we're used to in everyday life the difference is tiny – fractions of a fraction of a second – but the effect becomes stronger the faster you travel, and if you could approach the universal maximum speed limit – the speed of light – your experience of time passing would grind to a halt. Photons – the particles that make up all forms of electromagnetic radiation including light, microwaves, radio waves and X-rays – by definition move at the speed of light itself, and so they never experience any time passing at all, existing in an endless present.

For us the photons of the CMB are remarkable fossils, relics of an unimaginably ancient epoch when the universe was young, but from the point of view of the photons themselves any journey – even the longest trek in the universe – is over as soon as it has begun. They truly are a direct link to the origins of everything, and a link that you can see just by turning on an old-fashioned TV set.

This isn't the first time that bird droppings have played a pivotal role in astronomical legend. Apocryphally at least, our systems of measuring both time and position on the surface of the Earth also have their own strange link to avian incontinence.

The prime meridian line, defined by the axis of the Airy transit circle telescope at the Royal Observatory Greenwich, is the baseline for our system of measuring east–west position – zero degrees longitude. It is also the foundation for Greenwich Mean Time, which is based on the apparent position of the Sun over the prime meridian, and the world's time zone system, defined as hours ahead of or behind the time at Greenwich. Although the Greenwich telescope is no longer actively used for positional or timing measurements – which are now made much more accurately by orbiting satellites and atomic clocks respectively – still these modern measures are referred back to a notional line passing through the Royal Observatory.

And yet there is nothing particularly special about Greenwich itself. Unlike the equator, which is defined by the rotational axis of the Earth and forms a natural baseline against which to measure latitude, there is no equivalent natural feature to recommend itself as a baseline for longitude. Instead, the choice of a zero line for east–west position is an arbitrary one: all that's needed is an observatory to establish the necessary astronomical measurements and the agreement of everyone around the world to adopt the same line. That agreement came in 1884, after a month of heated discussions in Washington DC, and despite the objections of the French – who had a fine meridian line of their own passing through the Paris Observatory, which in many ways would have served just as well as the one through Greenwich. It's tempting to see the contest between Paris and Greenwich as a triumph for Britain in her age-old rivalry with France, but in this case the clinching argument was one of expedience rather than imperial might: after 100 years of navigational tables and charts provided by the Royal Observatory, the Greenwich meridian was already being used by about 70 per cent of global shipping and so it was simply the most convenient option.

But when the Observatory had been built 200 years earlier in 1675, the choice of Greenwich as a location was itself one of expedience. In the seventeenth century, Greenwich's claim to fame was royal rather than astronomical: situated on the south bank of the River Thames within easy reach of London it was the site of the royal palace of Placentia – a favourite home of the Tudor monarchs in the sixteenth century and the birthplace of both Henry VIII and his daughter Elizabeth I. On the hill above the palace was an old fortress, known as Duke Humphrey's Tower, which – according to rumour – had seen sterling service as a discreet lodging for Henry VIII's many mistresses. By 1675, the tower was in a sorry state but its foundations were still sound and, on land belonging to the Crown, they were a cheap and convenient piece of real estate. The story of how the Royal Observatory came to be built here has less to do with astronomy than it does with trade and politics, with the mistress of another king, Charles II, and – naturally – with incontinent birds.

In the seventeenth century, the finest scientific minds of Europe were gripped by a difficult challenge: the problem of finding longitude at sea. Far from the coast and out of sight of any visible landmarks, it's extremely difficult to know where on Earth you are. And yet this information is vitally important for almost every aspect of a maritime voyage: plotting a course, avoiding rocks, reefs and other dangers, and even estimating journey times so that you know how many provisions to carry. Sailors had known for centuries how to find their latitude, or distance north or south of the equator, by measuring the height of the Sun above the horizon at noon or the height of certain stars at night. But finding longitude, the distance east or west of your destination, was much more difficult to do and by the seventeenth century this had become an urgent problem as European nations vied with each other for power and influence, and the need to ensure the safety and efficiency of navies and trading fleets became ever more pressing.

Like other rulers across the continent, Britain's King Charles II was acutely aware of the problem, and of the advantages to trade and naval power that would be the prize for any nation that was able to solve it. But the catalyst that ultimately brought a solution came from one of Charles' many mistresses: the clever and spirited Louise de Kérouaille, Duchess of Portsmouth. Louise was originally from France but, as a leading light in fashionable London society and a favourite of the king himself, she wielded considerable influence at the English court. Perhaps from a sense of loyalty to her homeland – or else in return for lavish gifts and favours from Versailles – she regularly used her position to advance the interests of French visitors to the English capital. One of these was a young mathematician and amateur astronomer named St Pierre, who claimed that he had devised a way to solve the longitude problem using the motions of the Moon against the background stars – like Project Echo almost 300 years later, another example of an orbiting satellite being used for very practical purposes here on Earth.

Louise informed Charles, who set up a Royal Commission to assess St Pierre's idea. The answer was that it ought to work in principle, but that the necessary data – very detailed star charts and accurate predictions of the Moon's orbit – did not exist. In March 1675, at the recommendation of the Commission, Charles appointed the talented young astronomer John Flamsteed to be his 'Astronomical Observator', charged with the task of making the required observations. Lacking a dedicated observatory, Flamsteed was installed in the Tower of London to carry out his work. Stargazing might seem like an odd use for a building more usually associated with imprisonment and execution, and no one knows for sure exactly where Flamsteed set up his telescopes but, as the castle's highest point, the iconic White Tower seems like a good bet, and to this day the building's northeast turret is known as the Flamsteed Tower.

Flamsteed's tenure at the Tower is rather obscure and in the absence of detailed historical evidence a series of amusing legends has grown up linking him to the building's better-known residents, the famous ravens. They all centre around a conflict between the birds and the astronomer, who needed an uninterrupted view of the sky in order to carry out his duties. Some versions of the story claim that Flamsteed complained to the king because the ravens were disrupting his observations, either by constantly flying in front of the telescope or by incessantly croaking. However, in the most entertaining variant of the tale, it is the king himself who took offence at the birds, and in particular their habit of fouling the telescope lenses with their droppings.

Enraged by this disrespectful conduct, Charles is said to have ordered the birds to be removed – only to be reminded by his astronomer of the legend that, should the ravens ever leave the Tower, England and its monarchy would fall – which must have been a particularly sore point given that Charles had only regained his throne 15 years earlier. The king reconsidered his hasty words and decided that the ravens could stay after all, while Flamsteed would go to a new, purpose-built observatory beside his royal palace at Greenwich.

Thus it seems that the ravens of the Tower got off far more lightly than the poor pigeons of the Holmdel Horn Antenna three centuries later but, alas, unlike the story of the pigeons the raven's tale is almost certainly a myth. For centuries, the Tower of London was the traditional site of executions and so its association with these impressive carrion birds seems very natural but, in fact, historical research indicates that the ravens of the Tower are probably a recent addition, brought in during the late nineteenth century to enhance its gothic reputation. In any case, if birds were really such a serious problem, moving from the city to rural Greenwich would hardly have been a sensible solution. Today, the Royal Observatory is often mobbed by flocks of London's latest nuisance bird, the feral ring-necked parakeets, but even these colourful and noisy newcomers have failed to disrupt the Observatory's famous telescopes.

In June 1675, a warrant was signed for the creation of a Royal Observatory at Greenwich, and Flamsteed laid the foundation stone on the site of the old fortress in August of the same year, becoming the first in a long line of Astronomers Royal. Whatever the truth about the ravens' involvement, their story is so appealing that they and their droppings have earned an honorary place in the history of the prime meridian and Greenwich Mean Time. But perhaps the most important message we can take, both from the ravens of the Tower and the pigeons of Penzias and Wilson, is that the progress of astronomy is rarely smooth or predictable. Like every other field of human endeavour, astronomy is a product of its time, and sometimes the tiniest discrepancy or the most insignificant decision can turn out to have momentous consequences, changing our view of the cosmos in ways that nobody could have foreseen.

Water Everywhere

For a resident of the British Isles, surrounded by sea, shaded beneath a thick blanket of cloud and seemingly beset by constant rain, sleet and snow, it hardly needs saying that our planet is saturated with water. Liquid water covers 72 per cent of the planet's surface, and ice another 3 per cent. Water vapour makes up around 1 per cent of the atmosphere and even the rocks of the Earth's crust and mantle contain significant amounts of water, which plays an important role in plate tectonics and the movement of the continents. Without water, the Earth would be a very strange place: its geology, geography and climate would all be completely different.

As well as its planet-wide importance, water has a highly personal significance for human beings: after oxygen, it is the most immediate of our physical needs. People can generally survive for several weeks without food, but just a few days without water will usually prove fatal. Water makes up around 70 per cent of our bodies by mass, and so you could even say that we are largely creatures of water. Indeed, water seems to be an essential requirement for every form of life on Earth and its absence would therefore render our planet dead and barren. We only have to look at our two nearest planetary neighbours – bone-dry Venus and Mars – to see the difference that a lack of water can make.

It's hard to imagine our planet without this ubiquitous substance, and yet current evidence suggests that much of the Earth's complement of water couldn't have been present when our planet first formed. It must have arrived later on, from somewhere else in the Solar System. The water that surrounds us, in oceans, lakes, rivers and ponds – even the water that comes from our taps – is an alien substance, an exotic arrival from the depths of space.

Water itself is extremely common throughout the cosmos, although it usually takes the form of ice or a gas, rather than the wet, liquid state that is most familiar to us here on Earth. The constituent atoms that make up a molecule of water (H_2O) are hydrogen – by far the most abundant chemical element in the universe – and oxygen – which is much less common than hydrogen but still exists in vast quantities in the clouds of gas that float between the stars (for more on abundances *see* Made of Starstuff). Under the right conditions, wherever these two elements occur together there is the potential for water to form.

It is therefore hardly surprising that astronomers have detected water molecules in many different locations, from the icy moons of gas giant planets like Jupiter and Saturn to the atmospheres of dying stars and even the chaotic regions around supermassive black holes in the centres of distant galaxies. But even though water is such a common molecule out in space, there is still a great deal of mystery about how it arrived here on Earth. Over the last few decades, astronomers have tried to piece together the history of our planet's water, investigating potential sources elsewhere in the cosmos and trying to understand how this water could have subsequently made its way here.

One place where water has been found in particular abundance is in giant molecular clouds, the vast nebulae of dust and gas, often hundreds of light years across, in which new stars are born. Our Solar System would have formed in just such a cloud around 4.5 billion years ago.

By carefully observing how stars are being created today inside giant molecular clouds such as the Orion Nebula – a glowing patch of gas that is just visible to the naked eye below the three stars of Orion's Belt – we can piece together the details of our own origins in what must have been a very similar environment in the distant past. Within the cloud, gravity began to pull a dense clump of gas and dust together, becoming the seed for a new star: the Sun. Included alongside this collapsing gas would have been a large quantity of water molecules. But as the central part of the clump coalesced to form the infant Sun, radiation from this newborn star began to heat the surrounding gas, driving volatile substances like water away from the inner regions. By the time the Earth and the other inner planets began to take shape from the rocky debris orbiting close to the Sun, it seems likely that the material in this part of the Solar System would already have lost much of its original water content. So why is our planet so wet?

The question is a complex one, and scientists still don't know all of the answers. It is possible that at least some of the water in the original gas and dust cloud did survive the heat of the nearby Sun to be incorporated into the forming planet. Indeed, recent studies indicate that the crystalline surface of olivine – a common mineral component of interstellar dust grains – is very good at trapping and retaining water molecules, so some water could have been smuggled into the makeup of the Earth right from the start, wrapped around these tiny solid particles. Once the Earth had formed, chemical reactions deep inside the planet could have created more water by bringing hydrogen and oxygen atoms together. Volcanic activity would have then released the resulting water vapour from the depths of the Earth. But, to begin with at least, the newly formed Earth would not have had a protective atmosphere of the right temperature and pressure to retain liquid water, and water molecules reaching the surface would simply have evaporated straight back out into space.

It seems that a substantial fraction of the Earth's generous endowment of water must have arrived here at a later date and so to explain its origins we clearly need to look for alternative sources elsewhere.

Further out in the Solar System, where the heat of the Sun is less intense, there is certainly plenty of water available. Scientists who study the formation of the planets refer to the 'frost line', the distance from the Sun beyond which conditions are cold enough for water to remain stable as ice. In the early Solar System, this invisible boundary would have fallen somewhere between the planets Mars and Jupiter. Indeed, today we find that ice is still very abundant beyond this limit and many of the moons of the gas giant planets Jupiter, Saturn, Uranus and Neptune are made of a mixture of rock and ice. There is a great deal of water locked up in these moons, but they are bound in the powerful gravitational grip of their parent planets and so it is hard to imagine how they could ever escape and deliver their watery cargoes to Earth. So, for the source of our planet's water, we are again forced to look elsewhere.

Beyond the domain of the giant planets lie two further regions where water ice dominates. These are the Kuiper Belt, a ring of frozen debris at the edge of the Solar System, which includes the dwarf planets Pluto, Haumea and Makemake; and the Oort Cloud, a huge halo of icy fragments that surrounds the Solar System and possibly extends as far out as a light year from the Sun – almost a quarter of the distance to the nearest star, Proxima Centauri.

Most of the billions of objects that make up the Kuiper Belt and Oort Cloud are rather small – just a few kilometres across – so they are hard to study directly over such large distances. However, the largest objects here, such as Pluto and its fellow dwarf planets, are hundreds or even thousands of kilometres in size and appear through Earth-based telescopes as tiny, fuzzy dots – enough to give us some basic information about their temperature and composition and thus allow us to infer something about the vast number of smaller objects that share their region of space.

Observing from a distance of billions of kilometres, the Hubble Space Telescope has already been able to see that the surface of Pluto is frosted with a thick layer of frozen nitrogen, mixed with frozen methane and carbon monoxide. These three substances are also the main constituents of Pluto's thin atmosphere and it seems that they freeze out onto the surface during the decades-long Plutonian winter, then evaporate into a gaseous form again in summer. But Hubble has also been able to measure the density of Pluto's interior and it seems certain that it contains between 30 and 50 per cent water ice – and perhaps even a subsurface ocean of water if there is sufficient internal heat inside the tiny world to keep the water liquid.

A close-up view was obtained by NASA's New Horizons probe, which flew past Pluto on 14 July 2015 after a journey lasting almost ten years – the first spacecraft to rendezvous with an object beyond Neptune and the second to reach a dwarf planet. In order to complete its epic journey in a reasonable amount of time, New Horizons built up tremendous speed, using the powerful gravity of Jupiter to give it an additional boost on the way in what scientists refer to as a 'slingshot manoeuvre'. New Horizons' record-breaking velocity may have enabled it to reach Pluto in under a decade but such speed also brought its own set of difficulties: on reaching Pluto the spacecraft was travelling far too quickly to be captured by the dwarf planet's weak gravity and so it had to achieve all of its science goals in a matter of days as it flashed past Pluto and its moons at 14 kilometres per second – the success of its ten-year mission all depending on a perfect choreography of cameras and instruments.

In the final hours before closest approach New Horizons was programmed to temporarily break off contact with Earth in order to focus on an intense burst of scientific observations, leaving the mission scientists waiting anxiously for 22 hours. This might not sound like long compared to years that the spacecraft had spent on the drawing board even before its decade-long flight, but by the time it arrived at Pluto many members of the New

Horizons team had already committed significant portions of their careers to the project – all in the hope of a treasure trove of images and data. There was a small but real chance that the probe might collide with small fragments of orbiting dust and ice as it flew between Pluto and its moons and, travelling at such high speeds, even an impact with a dust-sized particle would have the potential to damage or even destroy the craft. Anxieties were heightened still further by a fundamental limiting factor of all space missions: the speed of light. After 22 hours of planned silence the scientists still had to wait 4.5 hours for the radio signal announcing that New Horizons had survived to travel the 4.8 billion kilometres between Pluto and Earth.

But their agonising wait was not in vain. On 15 July 2015, at 53 minutes past midnight Greenwich Mean Time, New Horizons re-established contact with Earth. Its mission had been a success and the probe began to download a vast database of images and data back to the eager scientists. Surprises came thick and fast. As a tiny world on the frigid margins of the Solar System it had been assumed that Pluto would be an ancient place, frozen and unchanged for billions of years. But nothing could be further from the truth. Pluto's terrain is characterized by extreme contrasts: light and dark, rough and smooth. The ground is sprinkled with exotic ices and snows, in the form of frozen nitrogen, methane and carbon monoxide, but it is frozen water that dominates here, forming the bedrock for the dwarf planet's geology in the form of vast flat plains and jagged mountains of ice several kilometres high. The sharpness of the mountains and the smoothness of the plains – which are conspicuously lacking in impact craters – all suggest that this is a young landscape, perhaps no more than 100 million years old.

This means that Pluto is still geologically active, but planetary scientists will now have their work cut out for them explaining how such a tiny world can continue to renew its surface 4.5 billion years after it first formed, erasing old craters and raising new mountains. The power source for all this activity remains a mystery. Pluto itself is tiny – just 2,380 kilometres across – and so it should long ago have given up any residual internal heat to space – unless its rocky core contains a far higher abundance of radioactive elements than the rocks of Earth. At this distance from the centre of the Solar System the Sun appears as little more than a bright star so the amount of warming it imparts to Pluto is extremely low. And Pluto is locked in such a tight gravitational embrace with its largest moon Charon that tidal heating of its interior seems unlikely. Whatever the answer, New Horizons has shown us that the ice at the edge of the Solar System is involved in far more dynamic processes than we had ever imagined. As it hurtles unstoppably onwards, it is hoped that the spacecraft can be repurposed to fly past another of the Kuiper Belt's larger objects – providing further insights into the frozen reservoir of the outer Solar System before it leaves the Sun's realm forever.

In the tale of Earth's water, the main significance of Pluto and its dwarf planet companions is that they are relatively large and easy to study, but in terms of the sheer number of icy objects at the edge of the Solar System, they are only the tip of the iceberg. It is the billions of smaller bodies occupying the Kuiper Belt and

Oort Cloud that provide a possible source for the water that fills our oceans.

Several kilometres across and composed of a mixture of ice and rock, these objects are sometimes described as 'dirty snowballs'. At such great distances, they are extremely difficult to study from our vantage point in the inner Solar System but from time to time something happens to bring one of them closer. The gravity of one of the planets, or even of a passing star, can occasionally dislodge these mountain-sized chunks from their distant orbits and send them tumbling towards the Sun.

When such an object approaches the warmer inner regions of the Solar System, heat from the Sun begins to evaporate the ice on its surface, sending jets of vapour and dust spraying out into space. This gas and dust forms a vast halo around the solid nucleus of the object and streams away under the influence of the radiation from the Sun, forming a long tail: the object has become a comet.

Comets can certainly make their presence felt. Although the comet's solid nucleus is only a few kilometres in size, the tail of gas and dust can stretch for millions of kilometres, becoming visible across the Solar System. Most comets speed through the inner Solar System in just a few months, looping around the Sun before heading back out into the depths of space on highly elongated, cigar-shaped orbits and only returning to our cosmic neighbourhood after decades or even centuries have passed. Some comets are even more elusive: after putting in a single appearance they are flung out of the Solar System entirely, never to return.

But not every comet survives a trip around the Sun. Each visit to the inner Solar System involves a small but real chance that the journey will end in an encounter with another member of the Sun's orbiting retinue of planets, moons and smaller objects. The evidence of such collisions is scattered all across the Solar System: wherever we find a solid surface – planets, moons and asteroids – it is peppered with craters, each one the mark of a violent impact event. Many of these craters were certainly caused by comets. In addition, we can be sure that, despite having no solid surfaces to record the scars, the gas giant planets have also suffered numerous cometary impacts over the history of the Solar System. Indeed in 1994 we were able to witness one, when dozens of fragments of Comet Shoemaker–Levy 9 collided with Jupiter, producing huge spreading blotches of sooty debris in the planet's upper atmosphere. Jupiter was struck again in 2009, demonstrating that such large-scale collisions are far more common than we had previously supposed.

So, could these cometary impacts, each one delivering billions of tons of ice, be the means by which water was brought to the parched surface of a newly formed Earth? Major comet collisions are relatively rare these days – the current rate certainly isn't enough to explain the abundance of water that we see around us here on Earth. But the Solar System wasn't always as quiet as it is today and its early history would have seen a much higher impact rate. Even once the main planet formation process was complete, the Solar System took a while to settle into its current configuration. There would have been a lot of icy and rocky debris still around and the potential for collisions was high.

There is even some evidence that the planets of the inner Solar System suffered a catastrophic bombardment between 4.2 and 3.8 billion years ago, perhaps associated with a major orbital rearrangement of the giant outer planets that scattered the icy objects at the edge of the Solar System in all directions. Based on dating estimates for impact-related lunar rocks and careful studies of cratering rates on the Moon and other bodies, this 'Late Heavy Bombardment' theory is still hotly debated, but either way it's clear that there would have been a lot more comets crashing into the Earth during this early part of its existence than there are today.

However, there's another twist in the tale of Earth's watery history. Comets certainly contain plenty of water, but is it the right kind of water? The hydrogen atoms inside a water molecule can come in different varieties, or isotopes: ordinary, 'light' hydrogen (H), and a rarer form of 'heavy' hydrogen known as 'deuterium' (D), which contains an extra neutron. (There's also a third isotope of hydrogen known as 'tritium', which contains two extra neutrons, but this form is unstable and rapidly undergoes radioactive decay, so it doesn't feature in this particular story.) Water molecules made with the deuterium isotope are around 10 per cent heavier than 'ordinary' H_2O molecules containing only the more common, light form of hydrogen.

The water in Earth's oceans contains a very characteristic mix of these two different hydrogen isotopes, with the light form of water outnumbering the heavy form by 3,200 to 1. So any candidate for the ultimate source of Earth's water will also have to match this chemical fingerprint. This is where the comet hypothesis starts to unravel slightly.

In the last few decades, several space missions have been mounted to comets and, alongside observations with powerful Earth-based telescopes, this has allowed detailed studies to be made of the chemical composition of a handful of comets. Appropriately enough the first such mission went to the most famous comet of all, when the European Space Agency's Giotto spacecraft flew past Halley's Comet in 1986. To the surprise of astronomers, Giotto's instruments showed that the heavy, deuterium-rich form of water is twice as abundant in the ice of Halley than it is in the oceans of Earth. Subsequent studies of Comets Hyakutake and Hale–Bopp also found that they contain double the amount of heavy water compared with Earth.

In 2014, the European Space Agency's Rosetta spacecraft went into orbit around the 4-kilometre-wide Comet 67P/Churyumov–Gerasimenko, becoming the first spacecraft to orbit one of these frozen hulks – no mean feat considering the extremely weak gravity of this mountain-sized object. During a ten-year journey, Rosetta had used a series of flybys of the Earth and Mars to build up the 15 kilometres per second required to match speeds with 67P. On the way, it had flown past two asteroids and made studies of the Martian surface and atmosphere, and so by the time it arrived at the comet it was already a seasoned science mission (ironically, as it flew past the Earth in November 2007 Rosetta had also been briefly misidentified as a potentially dangerous asteroid). Once at 67P, its cameras immediately began to map the comet's nucleus, revealing a bizarre, double-lobed structure like two mountains joined together by a narrow 'neck', the whole structure just 4 kilometres from end to end. Though small, the comet's surface was also surprisingly varied, consisting of vertiginous crags, rocky scree slopes and ashen-coloured plains dotted with pits, boulders and dunes of dust.

Rosetta also carried a suite of instruments to analyse the comet's composition from a distance, and a smaller landing craft, Philae, was dispatched to the surface to study it in situ. Although almost the size of a car, in the comet's weak gravity Philae weighed only around as much as a watch battery does here on Earth and, unable to find a secure purchase on the surface, it bounced twice before coming to rest several kilometres from its selected landing site when it wedged itself in a gully. Even in this undignified position, Philae was able to carry out its scientific mission, deploying instruments to 'taste' the cometary material and sending back a handful of close-up images of the comet's strange surface, where frozen water behaves in ways analogous to rock here on Earth – a type of geology never observed in detail before.

Rosetta's arrival was timed to coincide with the beginning of the comet's next journey to the inner Solar System, when the surface begins to warm and to emit gas and dust. As the comet neared perihelion – its closest approach to the Sun – in August 2015, Rosetta had a ringside view of the peculiar geological processes that reshape the surface of this tiny world every time it heats up. In particular it was able to observe the formation of the strange sinkhole-like structures that pit the comet as subsurface voids left by erupting vapour collapse in on themselves. Even tiny Philae had a surprise in store: as the levels of sunlight increased, the lander was able to recharge its batteries via solar panels and resume science observations from the surface itself.

A success on many levels, the Rosetta mission provided another opportunity to study one of the Solar System's reservoirs of frozen water in detail. The spacecraft's instruments sampled the water vapour coming off 67P's nucleus and found that, yet again, it failed to match that of the Earth, with up to three times as much deuterium as the water of our oceans. Comets are definitely very rich in ice, but it was starting to seem as though astronomers needed to find a different extraterrestrial source for our planet's water.

COMET CULTURE

Whatever the true extent of comets' contribution to the Earth's water, they have certainly made their mark on human culture over the centuries. The word 'comet' comes ultimately from Ancient Greek and means 'long-haired star' – a reference to the way a comet's tail appears to stream across the sky like waving tresses – and, as unpredictable intruders into the ordered regularity of the heavens, comets posed something of a puzzle for ancient philosophers. The Greeks assumed that they must be atmospheric phenomena since the heavens themselves were supposed to be perfect and unchanging. Meanwhile, ancient Chinese astronomers meticulously recorded and catalogued cometary appearances, searching for significance in the shapes and subtle variations of their tails.

In medieval Europe, comets were seen as omens of change – and to the medieval mind change usually meant something bad. A comet appears in the Bayeux Tapestry, just a few panels before the famous scene in which England's King Harold is shot in the eye with an arrow at the Battle of Hastings in 1066. This comet had appeared just a few months before the battle and, with the benefit of hindsight, the invading Normans were able to incorporate it into the story of their glorious victory – as an omen of doom for the perfidious English of course. In the tapestry the comet blazes ominously above the rooftops while figures on the ground anxiously point and debate its significance.

A comet is the herald of happier news in a fresco of *The Adoration of the Magi*, painted by the Italian artist Giotto di Bondone in the Scrovegni Chapel in Padua around 1305. Giotto depicts the Star of Bethlehem as a fiery comet hanging above the stable of the Nativity, perhaps inspired by a real cometary apparition that had occurred in 1301.

The Great Comet of 1577 caused a stir all across Europe and also found its way into paintings and engravings, but its most significant influence was philosophical rather artistic. The Danish astronomer Tycho Brahe compared observations of the comet made from Copenhagen and Prague and showed that, contrary to the ancient view of comets as meteorologi- cal phenomena, it must be further away from us than the Moon – well outside the Earth's atmosphere. Brahe's assistant, Johannes Kepler, took this iconoclasm a step further, pointing out that the comet's curving path should have sent it crashing through the crystal spheres that were supposed to hold the Moon and planets in place – surely a fatal blow for the traditional view of the cosmos as a system of celestial spheres centred on a stationary Earth? In the end it would take more than this to convince the intellectual establishment to abandon the idea that the

Earth was the centre of the universe, but Brahe and Kepler's work on the Great Comet started to chip away at the age-old certainties about our place in the cosmos.

The unruly reputation of comets was finally laid to rest in 1705 by Edmond Halley, who used Isaac Newton's newly formulated laws of gravity to show that a comet seen in 1682 followed a very similar path to comets recorded in 1607 and 1531. His conclusion was that they were all in fact the same object, which was moving around the Sun in a highly elliptical orbit lasting 76 years. Halley predicted that the comet would return again in 1758, which it duly did, although he didn't live to see it. However, the comet's return secured Halley's fame, in the process providing a neat test of Newton's physics and demonstrating that it isn't only planets that orbit the Sun. Using Halley's 76-year orbit and working backwards, it is clear that his comet was in fact the very same object that is immortalized in both the Bayeux Tapestry and Giotto's fresco. The European Space Agency's mission to Halley's Comet in 1986 – the first to photograph a comet's nucleus – was named Giotto to commemorate this connection between art and science.

By showing that comets obeyed the clockwork rules of Newtonian gravity Halley may have removed some of the mystery and fear associated with them but, rather than being tamed, comets quickly acquired whole new layers of meaning and significance.

As the eighteenth and nineteenth centuries progressed, science increasingly challenged old, cosy certainties about the natural world and our place within it, and by the middle of the nineteenth century advances in astronomy had revealed a universe so vast and so old that human lives and achievements seemed utterly insignificant by comparison. Comets, returning after decades or centuries travelling through the gulfs of inter-planetary space, became perfect symbols of the transience of human life in the face of cosmic spans of time.

In his Pre-Raphaelite painting *Pegwell Bay, Kent – a Recollection of October 5th 1858*, the artist William Dyce contrasted human, geological and astronomical timescales, showing a family of Victorian day-trippers beside the ancient chalk cliffs, while Donati's Comet hangs evocatively in the sky above. The same comet may have inspired writer Thomas Hardy to compose his poem 'The Comet At Yell'ham', in which the narrator reflects that both he and his lover will be long dead by the time of the comet's next visit.

Others took a more sensational approach, with pamphlets and news-paper articles that predicted death and disaster each time a new comet appeared. In contrast to the superstitions of previous centuries, these Victorian doom-mongers used (or misused) the latest science to lend credence to their stories. The discovery of traces of the poisonous gas cyanogen in cometary outflows led to hysterical claims that life on Earth could be extinguished if the planet passed through the tail of a comet – unless, of course, readers purchased a supply of protective 'comet pills'. (In fact, the Earth regularly passes through comet tails, with no more ill effects than a brilliant shower of meteors as comet dust burns up in our atmosphere. The cyanogen is far too thinly spread to pose a threat to living things.)

In the early twentieth century, the writer H.G. Wells turned this scenario on its head, imagining comets as extraterrestrial agents of social and political change. His 1906 novel, *In the Days of the Comet*, describes how gas from a comet's tail causes Earth's population to fall into a deep sleep. They awake to find all their violent and selfish instincts have vanished: 'The great Change has come for evermore, happiness and beauty are our atmosphere, there is peace on earth and good will to all men.'

Cometary reputations had changed again by the end of the twentieth century when they were implicated in the impact-related extinction of the dinosaurs, inspiring a host of comet-themed disaster stories in science fiction novels and movies, such as 1998's *Deep Impact*. But at the same time astronomers also began to suspect a more benign side to comets, as the possible source of the Earth's life-giving water and organic chemicals. Whether evil or benign, comets achieved something approaching celebrity status in November 2014, when the Rosetta spacecraft's lander Philae made a bumpy touchdown on Comet 67P, garnering global media coverage and even trending on the social media website Twitter.

Surprisingly, a potential solution to the mystery of Earth's water turned up much closer to home than anyone suspected. Apart from the Oort Cloud and the Kuiper Belt, the other main source of impacting objects in our Solar System is the asteroid belt, between the orbits of Mars and Jupiter. Like comets, asteroids are fragments of debris left over from the formation of the planets. They range in size from a few metres to hundreds of kilometres across but, unlike comets, they are mostly made of rock and metal rather than ice. In fact, because of their relative proximity to the Sun, astronomers had believed them to be unlikely sources of water – surely if the Earth struggled to capture and retain water from the original dust and gas of the solar nebula then the asteroids, with similar temperatures but weaker gravity, would have had an even harder task?

But this view changed when detailed studies of asteroids such as 24 Themis began to show that some of them do, in fact, harbour significant quantities of frozen water and – crucially – water of the correct isotopic ratio. Perhaps these icy asteroids originally formed just beyond the 'frost line' and were therefore able to acquire and retain significant amounts of water as the Solar System was forming? Or perhaps they migrated inwards from colder regions during the period of great planetary reorganization early in the history of the Solar System, a period when many of their siblings could have been sent hurtling towards the Earth?

Back on Earth itself, intensive study of a class of meteorites known as carbonaceous chondrites has shown that their mineral composition is also unexpectedly rich in water – again with an isotope ratio similar to that of our oceans. These ancient space rocks are also thought to originate in the outer parts of the asteroid belt, supporting the idea that it could harbour a watery reservoir – and that that water could easily have found its way to Earth.

This impression has been bolstered further by data beamed back from the asteroid belt itself by NASA's Dawn spacecraft, which visited the large asteroid Vesta in 2011, before moving on to rendezvous with Ceres in March 2015 (and beating New Horizons to the title of first mission to a dwarf planet by a matter of months). Even before Dawn's arrival, Ceres had been showing intriguing signs that it might not be quite as dry as the traditional picture of an asteroid would suggest. In 2014, the European Space Agency's Herschel Space Observatory (named, appropriately enough, for the discoverer of Uranus and the man who coined the term 'asteroid') reported faint signs of water vapour around the tiny world. Meanwhile, the best available images of Ceres were rather blurry shots taken by the Hubble Space Telescope, which showed a reddish-brown globe marked by a small but conspicuous white patch. As Dawn made its final approach, this patch resolved itself into two dazzling white spots, each just a few kilometres across and both lying in the bottom of a colossal 92-kilometre-wide crater.

The brightness of the spots is due to reflected sunlight, leading to speculation about their composition. Could it be that the giant impact that blasted out the crater millions of years ago had dug down below Ceres' surface layers of rock and rubble to reveal an underlying layer of ice? In this case, the dazzling white of the spots could either be due to ice itself or, more likely, to deposits of salt left behind when the exposed ice evaporated into the vacuum of space. Other, smaller craters on Ceres reinforce the idea of a subterranean icy layer: their edges have a softened appearance, almost as if they have slumped gently in on themselves – exactly as we might expect if their foundations lay on slushy ice rather than solid rock.

Along with studies of meteorites here on Earth, missions such as Dawn will doubtless continue to reveal new surprises about the watery secrets of the asteroid belt. But it may still be too early to completely dismiss comets from the running. Recent results from Comet LINEAR and Comet Hartley 2 indicate that, unlike Comet Halley, these two bodies seem to contain exactly the same ratio of heavy and light water as Earth. The crucial difference might lie in where exactly the comets have come from. Properties of the orbits of Hartley 2 and LINEAR indicate that the original home of these comets would have been in the Kuiper Belt, just beyond the orbit of Neptune, while Halley, Hyakutake and Hale–Bopp are all believed to have come from the more distant Oort Cloud. Perhaps the type of water carried by a comet depends on its place of origin?

The results from Comet 67P are less encouraging – it too is originally from the Kuiper Belt and yet, unlike LINEAR and Hartley 2, its water is very different from the Earth's. However, the complexity of 67P's surface structure makes it hard to draw firm conclusions about anything at this stage. The comet has clearly been heavily altered by its repeated close encounters with the Sun – perhaps the water vapour we currently see emerging from the nucleus is not entirely representative of 67P's original pristine composition.

Comets themselves may not be the only places where we can look for clues. The Earth's water might be hard to explain, but there is another world where the presence of water seems even more unlikely. At just one-third of the Earth's distance from the Sun and lacking all but the thinnest of atmospheres, tiny Mercury is blasted by ferocious solar radiation and has daytime temperatures in the hundreds of degrees. It is the last place anyone would think to look for water, and yet space probes have revealed the presence of significant quantities of ice, sheltered in permanently dark craters near the planet's poles. Since these craters are younger than the planet itself the water can't have been there since the formation of the Solar System – like the Earth's water it must have arrived later on. Once again, comets are prime suspects – an impression that has been reinforced by the discovery that the surprisingly dark colouration of Mercury's surface may be due to a dusting of carbon derived from comet dust.

Of course, the real problem is that we are trying to reconstruct events that took place in the distant past, using only the confused and ambiguous evidence that we find scattered across the Solar System today. It would be much easier if we could simply watch the process of planetary formation in action and wait to see how water is brought back to the warm inner planets. New telescopes may soon give us the opportunity to do just that, not by looking backwards into the history of our own Solar System, but by looking outwards towards planetary systems that are still in the process of formation elsewhere in the Milky Way. Already, around a handful of newly formed stars, such as Beta Pictoris, astronomers have found evidence for the presence of 'extra-solar comets' – icy bodies that originate in the outer reaches of these alien solar systems, producing huge tails of dust and gas as they swing close to the warmth of their parent star. Currently we infer the presence of these comets as their tails partially block the light of their star, but the next generation of powerful telescopes may allow us to study their properties in more detail and so ultimately the answer to the riddle of water's arrival here on Earth might come from further away than we ever imagined.

We encounter water every day in a huge variety of contexts and it is strange to think that the origins of such a familiar substance could be shrouded in so much mystery. Whatever the final answer, it seems likely that the water here on Earth reached us via a number of different routes and at various times in the first few hundred million years of our planet's history. But whether it arrived discretely, hidden away in the crevices of tiny dust grains, or violently in the explosive impact of comets and asteroids, much of the water in our oceans today has an alien origin, far off in the depths of our Solar System. The next time you take a shower or pour yourself a glass of water, it is worth remembering that those molecules of H_2O have a surprisingly exotic past.

Alien Oceans

Earth is a special planet. With a surface dominated by liquid water in the form of oceans, seas, lakes and rivers, it is unique in the Solar System. Its watery state is made possible by a combination of fortunate circumstances: our planet's orbit sits right in the middle of the Sun's Habitable Zone –also known as the 'Goldilocks Zone' where temperatures are neither too hot nor too cold but 'just right' for water to exist in a liquid state – and its surface is shielded from the vacuum of space by a protective atmosphere, which has just the right pressure to allow liquid water to be stable.

Although the Earth enjoys a privileged position in the Solar System, it might not be the only place where oceans exist. As we've seen, water is an abundant molecule throughout the cosmos and there are more ways than one to create the conditions under which it can be a liquid. Furthermore, water is not the only substance that can exist in a liquid state – so there is the potential out there for oceans that are extremely alien indeed, with properties very different from those we're used to here on Earth.

The Habitable Zone itself is a rather slippery concept and its boundaries are surprisingly hard to define. The further away from the Sun, the more thinly its radiation is spread through space, and so the less heat and light a planet receives per unit of surface area. In the simplest sense, the Sun's Habitable Zone can be thought of as the region of the Solar System that is far enough from the Sun to have a surface temperature of less than 100 degrees Celsius – so that water doesn't boil away – but near enough to the Sun to ensure that the surface remains above 0 degrees Celsius, preventing water from freezing solid. But this definition assumes that other conditions on a planet are also Earth-like: in particular it requires the planet to have an atmosphere that is similar in composition and density to that of our own. Under this definition, the Earth is currently the only planet that lies squarely within the Solar System's Habitable Zone, with Venus just slightly too close and Mars slightly too far from the Sun for comfort. But, as we'll see, if atmospheric conditions on Venus and Mars were somewhat different both worlds might still be able to support oceans of liquid water on their surfaces – and indeed they may well have done so in the distant past.

In any case, orbiting within the Habitable Zone is certainly no guarantee that a world will in fact be habitable. Factors such as day length and patterns of atmospheric circulation can have a huge effect on the distribution of heat around the globe: despite its favourable location in space, Earth still has polar ice caps where temperatures remain resolutely below 0 degrees Celsius for most of the year.

A more extreme example is the Moon, which is exactly the same distance from the Sun as the Earth and therefore shares its parent planet's prime position. And yet the Moon is barren and lifeless: the only 'seas' here are dark, smooth plains of solidified lava. These characteristic geological features, clearly visible from Earth, were once actually believed to be bodies of water and indeed they are still referred to as *maria* – Latin for 'seas'. Several of them bear romantically watery names: Mare Imbrium (the Sea of Showers); Mare Undarum (the Sea of Waves); Mare Spumans (the Foaming Sea). There is even an Ocean of Storms (Oceanus Procellarum), and numerous lunar features bear the title of 'bay' (*sinus*), 'lake' (*lacus*) and 'marsh' (*palus*). The justification for this maritime fixation was more to do with analogy than evidence. In 1609, Galileo's telescopic observations of the Moon revealed that it possessed uneven, mountainous landscapes that in many ways appeared to be like those of the Earth. But, alongside the rugged mountains, there were also dark, smooth areas. Although early telescopes didn't have the necessary magnification to reveal these features in detail, it seemed natural to assume that they too would have their counterparts here on Earth, and if the Moon had mountains, then should it not also have seas?

The Jesuit astronomer Roger Joseph Boscovich discovered evidence for the lack of a lunar atmosphere in 1753, but the belief that the Moon might be habitable, with the markings on its surface indicating oceans and vegetation, persisted into the early nineteenth century. William Herschel, the great eighteenth-century astronomer who had discovered Uranus, was convinced of the Moon's habitability, and his son John – himself one of the most famous astronomers of his day – was even unwittingly implicated in a newspaper scandal about lunar life. In what became known as the Great Moon Hoax of 1835, *The New York Sun* had published a series of sensational articles describing the flora and fauna of the Moon as allegedly observed by John Herschel through a powerful (and utterly fictitious) new telescope. The pieces were lavishly illustrated with lithographs of exotic landscapes and creatures, including winged, bare-breasted Moon maidens, and readers flocked to the newspaper, eager to believe in life elsewhere in the Solar System – especially if it was scantily clad. Needless to say, John Herschel had observed nothing of the kind, and he only later became aware that his name was being used in this way to sell newspapers. He responded with (mostly) good humour, quipping that his actual research could never hope to be as exciting, although it seems he eventually grew tired of explaining to people that the whole story had been made up.

By this stage, it was becoming abundantly clear that the Moon was airless and as dry as a bone, although the early lunar cartographers had been right in one sense: long ago the lunar 'seas' would indeed have been full of liquid, but liquid rock, not water. In the early days of the Solar System, when planets and moons were still being regularly pounded by rocky and icy debris in the form of asteroids and comets, large impacts would have been able to crack the crust of the Moon, allowing liquid magma to flow out and pool in the low-lying areas of the surface.

But these ancient lava oceans are the only seas the Moon will ever have. Without an atmosphere, liquid water cannot be stable here: instead it would quickly boil away into the vacuum or freeze into the soil. Air pressure plays a vital role in holding water in a liquid state, by pushing down on its surface and preventing its molecules from flying off in all directions. As well as providing sufficient pressure to keep water in a liquid form, an atmosphere provides another vital service, moderating the huge swings in temperature that an object would otherwise experience between day and night. On Earth, the highest temperature ever recorded is 70.7 degrees Celsius, in the deserts of Iran, while the lowest is -89.2 degrees, at Vostock Station in Antarctica. These are extremes, but for most of the time temperatures over the majority of the Earth's surface are far milder and stay within a much smaller range. By contrast, on the airless Moon, temperatures regularly swing between 123 degrees Celsius during the day and -233 degrees at night. Perhaps this was not always so: when the Moon first formed it may well have had a gaseous atmosphere. But the Moon is small and its gravity is weak: whatever air it might once have possessed has long ago escaped into space, and with it any chance of liquid water on the lunar surface.

In contrast to the Moon, the planets Venus and Mars are usually considered to be slightly too close and slightly too far from the Sun respectively to lie within the Habitable Zone – certainly neither of them has surface conditions that could support oceans today. But in fact, with the right type of atmosphere, it would be perfectly possible for both of them to maintain the temperatures and pressures necessary for water to be stable on their surfaces. Ironically, both planets have exactly the wrong kind of atmosphere to tip conditions in favour of liquid water. Venus is close to the Sun, receiving about twice as much solar radiation per square metre as the Earth, and unfortunately it also has a thick atmosphere of carbon dioxide which traps most of this heat, raising the surface temperature to around 450 degrees Celsius. Meanwhile, chilly Mars receives less than half the amount of solar radiation per square metre as the Earth and its thin atmosphere exacerbates the problem as it is inadequate either for trapping heat or for providing the necessary air pressure to keep water liquid over a sustained period.

However, there is strong evidence that in the distant past conditions on both these worlds were very different – enough to allow both of them to have abundant water and oceans of their own, despite their apparently unfavourable distances from the Sun. From relatively balmy initial states, both of our neighbouring planets seem to have undergone runaway changes that drove them further and further from the conditions required to support liquid water – eventually resulting in the total loss of their oceans. Meanwhile, over the same 4.5 billion-year period, the Earth's atmospheric, geological and biological systems have conspired to keep surface conditions within the critical range, even though the Sun's output of heat and light has actually increased by around 25 per cent during this time.

On Venus, it was perhaps its proximity to the gradually brightening Sun, along with its lack of a protective magnetic field, that was the planet's undoing. Water vapour is a powerful greenhouse gas, so as it evaporated from Venus' oceans it would trap more heat, warming the planet still further. In addition, the Sun's savage ultraviolet rays are able to split water molecules apart, dissociating them into hydrogen atoms and hydroxide molecules. Without a magnetic field to protect the Venusian atmosphere from the full force of the solar wind, the extremely light hydrogen atoms would be swept from the planet's gravitational grip and lost into space. So, unlike on Earth, where a cycle of evaporation, cloud formation and rain continually removes then returns water to the oceans, on Venus evaporation was a one-way street – and the Venusian oceans gradually dwindled and disappeared. Today, Venus' highly reflective cloud deck, which completely shrouds the surface and ensures that the planet is the brightest object in our night sky after the Moon, is composed not of water but of droplets of concentrated sulphuric acid. But, as with the Moon, the default assumption of early astronomers was that Venus would have oceans, leading to speculation that the planet's clouds concealed an exotic landscape of seas and swamps, teeming with prehistoric creatures – an idea that persisted well into the twentieth century, when space probes revealed the dry and hellish truth about the Venusian climate.

Just beyond the outer edge of the Habitable Zone lies Mars and, like the Moon, the Red Planet's fate was probably sealed by its small mass and correspondingly low gravity – just a third that of the Earth's. Like the Moon and Venus, Mars also lacks a global magnetic field, leaving its atmosphere to bear the full brunt of the solar wind. With just half the diameter and a tenth the mass of the Earth, the planet's grip on its atmosphere has always been weak, and over billions of years much of it has been eroded away by the continual bombardment of particles from the Sun. The process is still ongoing: NASA's Mars Atmosphere and Volatile EvolutioN mission (MAVEN) probe, which went into orbit around Mars in 2014, has sent back data showing a trail of atmospheric gases streaming behind the planet like the tail of a comet at a rate of around 1 tonne per hour.

Today, the Martian atmosphere is composed mostly of carbon dioxide but the heat-trapping greenhouse properties of this gas are severely limited by the fact that the atmosphere is extremely thin – air pressure at the surface is less than 1 per cent of the sea level value here on Earth. This means that both the temperature and pressure on Mars are now too low for water to exist in a liquid state – at least in the open air. If you placed a bucket of water on the surface of Mars, it would be a race between evaporating into the thin atmosphere and freezing solid in the intense cold. But extrapolating the current atmospheric loss measured by MAVEN backwards in time suggests that in the past Mars would have been a very different place. Around 3.8 billion years ago, it may even have possessed an atmosphere thick enough and warm enough for liquid water to have been present in large quantities.

Over the last few decades a fleet of orbiting spacecraft and surface probes and rovers have sent back tantalizing evidence that, in the distant past, the surface of Mars was indeed a very wet place. Space-based images reveal dried-out river deltas and winding canyons, perhaps carved out by catastrophic floods, and signs of ancient shorelines fringing desiccated lakes and seas. Down on the ground the rover missions have found evidence of pebble-strewn riverbeds, layers of sedimentary rock laid down by water and minerals such as clay and haematite, which form in wet conditions.

In fact, physical evidence for Mars' warm and wet past had been sitting unrecognized in museums and laboratories here on Earth for half a century before the first space probe ever reached the Red Planet. The story of how these clues reached our planet without the aid of rocket technology, and how scientists finally came to recognize them, reveals a great deal about our hopes and preconceptions about the habitability of the other planets of the Solar System. It also demonstrates a surprising degree of physical contact between the Earth and its neighbouring worlds – an exchange that has been taking place for billions of years.

This fact was perhaps never clearer than on 28 June 1911 when, at 9am local time, a visitor from space streaked across the clear blue desert sky of Egypt. Travelling at several kilometres per second, this alien traveller was a piece of rock weighing more than 10 kilograms. As it plunged through the air, friction heated its surface, causing it to melt and blacken. At last, the stresses and strains of the chaotic descent became too much and over a region of lush vegetation, criss-crossed by irrigation canals, the rock exploded with a series of loud detonations, shattering into fragments that buried themselves in the ground below. When the dust had cleared an 11-million-year journey was finally over.

This dramatic event was not without witnesses. The fragments had fallen in the Nile Delta, in the farming district of El Nakhla el Baharia, and it wasn't long before news had reached the local press. Newspaper reports described a 'fearful column' of white smoke, a 'terrific noise' and a shower of 'black stones'.

Within a few days, the Nakhla region received another foreign visitor, this time in the form of W.F. Hume, the British director of the Geological Survey of Egypt. Hume immediately realized that the newspaper accounts described a meteorite, a rock from space that had survived its violent descent through the Earth's atmosphere. Unlike the more common meteors, which burn up entirely high in the air, meteorites are those slightly larger objects that make it all the way down to the ground. They are very rare and, as precious samples of material from beyond the Earth, also extremely valuable.

In an article published in the *Cairo Scientific Journal* in August that year, Hume wrote:

> On the clear cloudless nights which we so often enjoy in Egypt, interest is frequently aroused by a brilliant 'shooting star' or brightly-coloured meteor shooting across the darkened vault of the sky. But of far deeper significance are the messengers from the vast extra-terrestrial regions of space which reach the earth, possibly witnessing to events of a far-distant past. Egypt has at length been honoured by one of these rare visitors.

But even Hume couldn't have guessed quite how rare, or how significant, the meteorite that fell over Nakhla would turn out to be. Like any good scientist, he immediately set about gathering as much evidence as he could. The stones were scattered over several square kilometres of farmland, and many had buried themselves up to 20 centimetres deep in the ground. Eyewitness accounts were vital for piecing together the chain of events and for tracking down precious samples of the meteorite itself, so Hume relied heavily on the detailed accounts of local people.

About 40 fragments were recovered altogether from Nakhla and surrounding villages, and the total haul amounted to over 10 kilograms of precious space rock. It seems likely that more could have fallen unobserved between settlements, where they may still lie to this day. Considering their exotic origins, the final resting places of the Nakhla fragments do seem rather prosaic, but most were only found because people working in the fields had actually seen them land: one near a water wheel, another in a cotton plantation, and one – rather unglamorously after such an epic journey – beside a compost heap. The original newspaper report even claimed that one piece had landed on a dog leaving the poor animal 'like ashes in a moment', although this story is almost certainly apocryphal (*see* Invaders from Mars, page 87).

The samples from Nakhla are beautiful rocks. The outside of the stones has a dark, silky appearance that Hume described as being like 'a glistening black varnish'. This is the 'fusion crust', where the outer layer of the meteorite has melted in the heat of the descent and then re-solidified. Inside, the rock is a delicate greenish-grey, glinting with tiny crystals and quite unlike the majority of other meteorites.

Normal meteorites are either fragments of shattered asteroids or leftover clumps of the primitive material from which the Sun and planets were built (*see* Sky Metal, page 293). Either way, they are very ancient, dating from the formation of the Solar System, and have barely changed in 4.5 billion years. This makes them very different from most earthly rocks, which are relatively young, created in the constant volcanic upheavals of our planet's crust, then moulded and transformed by geological process. Unusually for a meteorite, the Nakhla stones also do not appear to be very ancient, and the appearance of the rocks with their tiny crystalline grains seems suspiciously volcanic in nature. But how could that be, when they have clearly fallen from space?

Hume himself was well aware of this strange contradiction. Noting that Nakhla's unusual properties fell outside the normal meteorite classification system, he remarked that the stones nevertheless had 'close resemblances to certain rocks on the earth's surface'. But while he had identified this important clue, the mystery of the meteorite's true origins would not be solved for more than 70 years.

Museums around the world soon received pieces of the meteorite as gifts from the Egyptian government and, very quickly, the precious stones were sliced for scrutiny under the microscope and ground into powder for chemical analysis. But all to no avail: the leading meteorite experts of the day agreed that the Nakhla rocks were unusual but, despite detailed investigations, the fragments stubbornly refused to reveal their secrets.

For decades the glossy black stones with their pale green interiors remained on laboratory benches or locked away in museum cabinets. Where had they come from and how had they formed? Although answers were not forthcoming, the Nakhla stones gradually came to be linked with a handful of other unusual space rocks whose properties also couldn't be reconciled with standard classifications. Like Nakhla, these meteoritic oddballs were named after the places where they were found: Shergotty (now Shergahti) in India, Chassigny in France, and the improbable-sounding Elephant Moraine (a glacial feature in Antarctica). By the 1970s, it was clear that all these rocks were young, about a billion years old, compared with the 4.5-billion-year ages of most other meteorites. Also, like Nakhla, they had a distinctive volcanic appearance, suggesting that they'd formed from magma cooling deep beneath a planetary surface. Together, these properties seemed to rule out an origin among the asteroids, the ancient fragments of space rubble that are the source of the vast majority of meteorites.

Despite Earth's abundant volcanoes, chemical analysis clearly showed that the meteorites also had a very different composition from terrestrial volcanic rocks. Nor did they match the samples returned from the Moon by the Apollo missions. But there was one other candidate: a planet where space probes had recently revealed some of the largest volcanoes in the Solar System – Mars.

INVADERS FROM MARS

Although the idea of rocks being blasted from one planet to another may sound far-fetched, by far the least believable aspect of the Nakhla meteorite's story is the claim, made in the original newspaper report in 1911, that one of the fragments landed on a dog, leaving the unlucky animal 'like ashes in a moment'. If the account of the dog's demise is true, the poor animal could well be considered the first Earthling to be killed by a Martian, but despite being one of the most famous aspects of the Nakhla saga, there are a number of elements to the dog's story that simply don't add up.

The meteorite fell on the morning of 28 June, yet the newspaper gives the date of the dog's fatal encounter as 29 June. Moreover, the village where the dog was supposed to have been killed was 33 kilometres from

El Nakhla el Baharia. The sound of the explosion had been heard clearly, even this far away, but no meteorite fragments were ever recovered from the region. Even if they had been, there is another problem with the story: the alleged manner of the victim's death. Despite their glossy black crust of melted rock, the stones found at Nakhla were cool to the touch, even shortly after they'd landed. This is normal for all but the smallest meteorites; after drifting through the frozen depths of space for millions of years they are, quite naturally, chilled to the core. The heat produced by their rapid descent through the atmosphere is enough to melt the exterior but this stage of their journey is over in minutes and the heat may not have time to penetrate to the centre. Once the meteorite has slowed sufficiently the heating stops and the frozen interior rapidly reasserts itself, solidifying the surface and quenching any remaining heat. Incineration therefore seems an unlikely consequence of being hit by one of them. A head-on encounter with a meteorite is certainly not something to contemplate lightly but blunt force would probably be the main cause of concern for anyone on the receiving end.

A year after the meteorite fell, an official report dismissed the story of the dog as 'doubtless the product of a lively imagination' and the consensus among scientists now is that the tale was invented to spice up the story. Given what we now know about the origins of the Nakhla stones, it hardly seems necessary to embellish their extraordinary tale. As samples of the geology of another planet, they are of incalculable scientific value: as well as helping us to understand the watery past of an alien world and its potential as a habitat for primitive life, they may also be telling us something profound about our own origins here on Earth.

A vast number of meteorites fall to Earth each year, but most go unseen and undiscovered. Of the 61,000 that have been collected over the last few centuries almost all are rocky and metallic fragments of shattered asteroids or leftover debris from the very earliest days of the Solar System. These meteorites give us invaluable insights into the initial conditions and raw materials from which the Solar System formed, as well as a record of the first stages of the planet-building process itself, as rocks collided and merged together to form progressively larger objects. But these rather primitive meteorites can't tell us about more recent conditions on fully formed planets, and this is where rocks like Nakhla and the other Martian meteorites come into their own.

About 130 confirmed Martian meteorites have so far been identified and work is ongoing to pinpoint their regions of origin on the Martian surface and to estimate in which period of Mars' long geological history each rock was formed. But Mars rocks are not the only pieces of another world to have been blasted into space from their parent body and ultimately find their way to Earth. Between 1969 and 1972, NASA's Apollo missions brought back around 380 kilograms of lunar rock and dust to Earth, while later in the 1970s three unpiloted Soviet craft also returned a few grams of lunar material.

But it was only in the1980s that scientists realized that rocks with identical chemistry to the precious Apollo samples had been on Earth all along. Around 120 of these lunar meteorites have now been identified in museum collections, forming a useful (and free) supplement to the rocks that were artificially transported from the Moon to the Earth by the astronauts at such vast expense. Like their Martian counterparts, the lunar meteorites presumably arrived here in the same way, being blasted off the surface of the Moon by large impacts, then finding their way to Earth – a much shorter journey in this case since they were already orbiting the Earth in the first place as part of the Moon. They provide valuable first-hand evidence of the geological history of another world but, like the Martian rocks, the violent nature of their departure, their long journey through the harsh environment of space, and the rigours of their arrival on our planet mean that they require careful analysis and interpretation in order to distinguish their original characteristics from any changes that they may have undergone en route to Earth.

The existence of these extraterrestrial stones itself begs a very interesting question: if Moon rocks and Mars rocks have made their way here by natural processes, could rocks from the Earth have made the reverse journey? Earth's thick atmosphere and higher gravity mean that only a very large and violent impact would pack the necessary punch to blast Earth rocks all the way into space, but such impacts have undoubtedly occurred many times during our planet's 4.5-billion-year history. Computer simulations support the idea that rocks from our planet could make their way to many distant corners of the Solar System, so it is perfectly conceivable that samples of terrestrial geology could be casually lying even now on the decidedly extraterrestrial surface of an alien world.

This is a tantalizing prospect for palaeontologists studying the evolution and origins of life on Earth. Plate tectonics and other forms of geological activity mean that rocks older than a few hundred million years are quite rare in the Earth's crust. Most ancient rocks have been eroded away by wind and rain or subducted back into the planet's interior. This is highly inconvenient for scientists trying to piece together the history of life on our planet since it is precisely these ancient rocks that might contain a fossil record of the earliest life forms. But perhaps they've been looking in the wrong place all along. If impacts have been periodically sending consignments of Earth rocks sailing into the sky for the last 4.5 billion years, then the precious fossils that the palaeontologists seek might not be lying beneath their feet here on Earth but instead high above their heads on the Moon, preserved for billions of years on its barren, airless and geologically inert surface. The clues to our planet's past – and our own origins – could be out there in space. It is intriguing to think that future lunar expeditions might include palaeontologists as well as geologists and physicists, and that one day the most valuable rock samples returned from the Moon might actually have originated on Earth in the first place.

In fact, the search inside meteorites for tiny fossils from this crucial early period of geological history is already underway – but the ancient meteorites in question are from Mars, not Earth. The Martian meteorites have already told us unequivocally that billions of years ago Mars was a warm, wet world, with conditions – and perhaps oceans – not unlike those of our own planet. If this was the time when life was first gaining a foothold on Earth could it be possible that life was also getting started in the similar conditions of Mars? If so, could fossil evidence of these early Martians be preserved in the rocks that have found their way to us?

One Martian meteorite has provoked excitement and controversy in equal measure: Allan Hills 84001, named after the hilly region of

Antarctica in which it was discovered in 1984, is rare and unusual even among the other Martian meteorites. Dated at around 4 billion years old, from the period when Mars would have been at its wettest, the 2-kilogram rock made headlines in 1996 when US President Bill Clinton announced on national television that scientists might have found evidence for tiny fossil organisms hidden deep inside the stone. Scanning electron micrograph images showed small tubular structures that resembled tiny bacteria, and nearby traces of organic chemicals hinted that these features might once have been associated with biological activity. Since the original claims were made, ALH84001 has been subjected to intense scientific scrutiny, and problems with the initial interpretation quickly emerged. Perhaps the most serious was that the tubular 'fossils' were just a few tens of nanometres in size, far smaller than any known terrestrial bacteria and arguably too small to contain all of the biochemical machinery necessary to sustain life. Meanwhile, the organic traces found inside the meteorite could be due to terrestrial contamination that occurred after it arrived on Earth. The current consensus is that the features inside ALH84001 are not fossil lifeforms after all, but it certainly seems clear that the rock had a long association with liquid water during its time on Mars and the level of interest generated by the 1996 claim demonstrates that the possibility of Martian fossils is taken very seriously indeed.

If scientists consider it feasible for fossils to make their way from the Earth to the Moon or even from Mars to the Earth, is it conceivable that living microbes could survive those same journeys, sheltering in crevices deep inside a wandering piece of impact debris as it shuttles between worlds? For a long time most scientists would have answered this question with a resounding 'no'. After all, to survive such an interplanetary trip any microbes would first have to endure the heat and shock of the initial explosion followed by millions of years in the airless cold of space, where they would be constantly bombarded by harsh radiation. Finally, they would be subjected to another burst of heat and concussion as they reached their destination. Surely even the hardiest of bacteria would be destroyed by these rigours?

But life seems to have an endless capacity for surprising us, and in recent years evidence has emerged that makes the possibility of planet-hopping microbes seem closer to science fact than science fiction. We now know that some bacteria here on Earth are capable of surviving in a dormant state for extremely long periods, perhaps even millions of years, waiting for conditions to become favourable for them to emerge and begin to reproduce. Meanwhile, the study of 'extremophiles' – microbes that survive and even thrive in extremes of temperature, pressure, acidity or alkalinity – has shown that the range of conditions under which terrestrial life can exist is much wider than we had previously imagined. Some species of bacteria even live quite comfortably in the high radiation environments around nuclear reactors. Taking all of these different adaptations together, it seems that the biological toolkit for surviving an interplanetary trip already exists, albeit scattered among a range of microbial species.

Space agencies at least are taking the possibility of hitchhiking micro-organisms very seriously. In the early days of Solar System exploration, it was assumed that the vacuum and radiation of space would sterilize interplanetary spacecraft during their journeys, but now great care is taken to deep-clean craft and instruments before they are launched to avoid potential contamination. After all, it would be ironic if a mission sent to look for life on Mars simply ended up detecting life that it had brought with it from Earth. Even at the end of their missions, spacecraft are now disposed of in ways that minimize the possibility of them contaminating distant worlds. In September 2003, after spending eight years exploring Jupiter and its satellites, NASA's Galileo probe was commanded to end its tour by plunging into the atmosphere of the giant planet at over 170,000 kilometres per hour, incinerating itself and removing any possibility that it might accidentally seed one of Jupiter's icy moons with terrestrial bacteria. A similar fate awaits the Cassini spacecraft when it finally completes its mission at Saturn in 2017.

Back here on Earth, biologists have long understood that plants and animals often colonize remote ocean islands by being swept out to sea on floating rafts of vegetation. It is intriguing to think that meteorites might have played a similar role carrying life across the vast gulfs of interplanetary space. If, just once in the 4-billion-year history of life on our planet, bacteria have managed to survive the journey from Earth to Mars aboard a meteorite then any fossil Martians we might one day discover could be more closely related to us than we expect.

There is one final twist to this tale of planetary colonization. Mars is smaller than the Earth and its gravity is weaker, making it easier for Martian rocks to be blasted into space, and requiring less brute force to do so. Mars is also further from the Sun, so it is easier for Martian rocks to fall inwards towards the Earth than it is for Earth rocks to travel outwards to Mars. Over the history of the Solar System, we might therefore expect more inbound traffic to Earth than outbound traffic to Mars – which begs a disturbing question. If life began on Mars when it was a warm, wet world and then hitched a ride to the similar conditions on neighbouring Earth, then the Martians might already be here. They could be us.

When the Nakhla meteorite landed in 1911, the Red Planet had already been a target for observation, as well as speculation, for many decades. Through their telescopes, nineteenth-century astronomers had measured Mars' size and day length, and watched seasonal changes unfold as clouds of dust and glinting polar ice caps waxed and waned across the planet's surface. It almost seemed as though Mars might not be too dissimilar to the Earth – a dry, cold desert world to be sure, but maybe one that was still capable of supporting life – and perhaps even a techno-logical civilization. Chief proponent of the idea of civilized Martians was American astronomer Percival Lowell who, through his telescope in the desert of Arizona, thought that he could see a vast network of canals criss-crossing the Martian surface. Just as the farmers of the Nile Delta used canals to bring life to the Egyptian desert, Lowell reasoned that intelligent beings on Mars had constructed these great waterworks to irrigate the sands of their arid world.

Lowell's ideas were widely circulated and proved particularly popular with writers of science fiction stories. Classic tales by H.G. Wells, Edgar Rice Burroughs and Ray Bradbury can all trace their roots back to Lowell's exotic vision of ancient Martian civilizations, and these romantic views of the Red Planet still colour our imaginations today. But by the early twentieth century, scientific support for an inhabited Mars of canals and advanced lifeforms was already fading. As techniques of observation improved, it became clear that Mars was not simply a redder, drier version of Earth. Even at the equator, ambient temperatures were well below freezing, and the rarefied atmosphere was little more than a wispy halo of carbon dioxide, barely cloaking the planet. Apart from the frozen ice caps of the polar regions, there was no sign of any liquid water on the surface. Damning for Lowell, the Martian canals turned out to be nothing more than an optical illusion, conjured up by the human brain as the telescopic image of Mars shimmered and blurred with the effects of turbulence in the Earth's atmosphere. As the twentieth century advanced, a picture emerged of Mars as a dry, dead planet.

When NASA's robotic Mariner and Viking spacecraft arrived at the Red Planet in the 1960s and 1970s, they found exactly what everyone was already expecting: a dry, frozen world devoid of animal or plant life, and quite definitely unmarked by canals or other works of civil engineering. But as old, romantic ideas of Mars were finally laid to rest, a new and exciting vision of the planet was about to emerge, as the close-up images sent back by these spacecraft also contained a host of surprises.

Giant volcanoes dotted the dusty landscape, the largest of which, Olympus Mons, was three times higher than Mount Everest. And, despite the obvious lack of liquid water and the absence of

Lowell's artificial canals, in places the Martian terrain was marked by networks of ancient channels, which looked very much like the dried-up beds of river systems on Earth. If Mars was now cold, dry and dead, could it once have been warmer, wetter and perhaps, after all, hospitable to life? The question was tantalizing, but impossible to answer. To confirm the presence of flowing water in Mars' ancient past would require detailed studies of its rocks and minerals, and that would require Martian rocks to be returned to laboratories on Earth – a task well beyond the capabilities of spacecraft of the time.

Although rock samples couldn't be sent back, huge amounts of data could, and buried in this stream of information were the long-awaited keys to the origins of Nakhla and the other unclassifiable meteorites. Measurements of chemical abundances carried out by spacecraft on the surface of Mars were enough to provide a positive match for the enigmatic meteorite samples back on Earth. What's more, tiny bubbles of gas, trapped inside the meteorites as they solidified from magma, were identical to the gases of the Martian atmosphere. After 70 years, the secret of the Nakhla meteorite was finally out: from the sands of Mars to the deserts of Africa, it was a messenger from another world – and a priceless geological sample of another planet in the Solar System.

Now the scientific tables were turned. If data from Mars had solved the longstanding mystery of Nakhla, could data from Nakhla help to solve the brand-new mysteries of Mars? After all, nature had already achieved what spacecraft could not: samples of Martian rock were already on Earth, sitting in museums and laboratories around the world and waiting to reveal the geological history of the Red Planet.

With their origins revealed, the strange characteristics of Nakhla and the other Martian meteorites were at last making sense. Further lab studies began to piece together the history of the rocks, using subtle chemical and structural clues to construct a detailed biography from their formation to their arrival on Earth.

Solidifying from magma deep beneath the surface of Mars around 1.3 billion years ago, the rocks that would one day end up in Egypt led an uneventful existence for millions of years. Then, about 600 million years ago, a large asteroid or comet crashed on the surface of Mars, in one of the many violent collisions that punctuate the history of the Red Planet. Although some distance from the Nakhla rocks, the crash had two effects. Shockwaves passing through the ground fractured the rocks, leaving a network of tiny cracks. Shortly afterwards the rocks were flushed with liquid water, perhaps as ice deposits in the soil were melted by the heat of the impact. As the water receded the cracks were filled with deposits of clay, a mineral whose presence is a telltale indicator that liquid water must once have been present.

All that remained for the rocks was to somehow transport their precious evidence to Earth, but for that they had to wait another 600 million years. Another giant impact rocked Mars, but this time the Nakhla rocks were caught directly in the blast and hurled space-wards. Such was the force of the impact that they were blasted completely off the planet, becoming part of the free-floating flotsam and jetsam of the Solar System. But their journey was not over yet – another 11 million years of drifting was required before the Earth's gravity could capture them and draw them down to their final incandescent plunge over North Africa. The last 100 years spent in museum labs, being ground, sliced and etched by eager scientists is simply the latest, very short, footnote in an extremely long story.

The presence of rocks that formed in a wet environment indicates that the Mars of the distant past was very different from the cold, dry planet that we see today: a warmer world with a thick blanket of air and perhaps even with rivers, lakes and seas of liquid water. In short, conditions very similar to those on the Earth when life was first getting started here. Even if Mars is now cold and barren, it now seems at least possible that primitive life could also have existed there billions of years ago.

If evidence of such ancient Martian life could be found, it would profoundly change the way we think about ourselves and our place in the universe. If life can evolve twice in the same solar system, even in such a marginal environment as Mars, then surely it must be common throughout the galaxy – perhaps even on one of the hundreds of as-yet unexplored planets, recently discovered around other stars.

Finding signs of life on Mars will not be easy but, heeding the message of the Nakhla meteorite and its Martian cousins, the mantra of NASA's current fleet of spacecraft is 'Follow the Water!' An exciting new era in Mars exploration has now begun with probes and rovers scanning the planet for more rocks that formed in the presence of liquid water millions of years ago. The stakes are high: the deep history of Mars could well hold clues to the origins of life here on Earth.

On 6 August 2012, at 5am Greenwich Mean Time, another visitor from space streaked across another desert sky. This object weighed more than a tonne and it had been travelling for a little over eight months. The sky it traversed was pink, not blue, and canals and vegetation were conspicuously absent from the rust-coloured landscape of Mars. And, unlike the chaotic fall of the Nakhla rock just over a century before, this descent was a precisely choreographed exercise in technological bravado.

Having slowed the object from its cruising speed of 5.8 kilometres per second, a heat shield was jettisoned – no melted fusion crust here – and a parachute deployed to grasp at the thin air. At a height of 1.8 kilometres the parachute was released and eight downward-pointing rockets ignited, slowing the descent still further until the object hovered just metres above the surface. If the descent itself had been risky, the most daring part of the operation was still to come. From the hovering platform, nylon cables began to unspool, carefully lowering a payload towards the ground. This surreal contraption is known as a sky crane, and nothing quite like it had ever been used before. The creators of the sky crane couldn't be certain that it would actually work, even in gravity just a third as strong as the Earth's.

But it did: the payload touched gently down and the cables were cut. Its task now complete, the sky crane flared its rockets once more and, in a final act of self-sacrifice, hurled itself through the air, crashing down a kilometre away from the precious payload. Seven hectic minutes after hitting the top of the atmosphere, NASA's Mars Science Laboratory had arrived on the Red Planet. Standing on the red ground on six sturdy wheels, it waited to receive its first commands from Earth.

The size of a jeep, Mars Science Laboratory (MSL) is the largest and most ambitious spacecraft ever to touch down on another planet, and comes complete with cameras, robot arms and a geologist's toolkit for analysing the rocks of this distant world. Inspired by meteorites like Nakhla, MSL's mission is to unravel the ancient history of Mars as told through its geology, to search for signs of past and present water and, most ambitiously of all, to assess whether conditions suitable for life might once have existed on the Red Planet. As it works its way through the geological record, the rover, also known as Curiosity, has already found evidence for abundant water in the deep Martian past. Gravel deposits mark the course of ancient streams, and layers of sedimentary rocks, laid down in shallow lakes or slow-moving rivers, testify to the presence of extensive bodies of standing water.

Whether or not reservoirs of liquid water still linger deep beneath the sands of Mars, the case for underground oceans elsewhere in the Solar System is extremely strong. But, in a twist that would have pleased the lunar cartographers of the seventeenth century, these extraterrestrial seas are not on any of the planets, but on several of their moons. Perhaps the most exciting prospect is Jupiter's moon Europa, one of the four Jovian satellites discovered by Galileo just months after his first telescopic observations of our own Moon. For more than three centuries after its discovery, Europa was just a tiny point of light in the eyepieces of Earth-based telescopes, and the first ever close-up images sent back by NASA's two Pioneer spacecraft in 1973 and 1974 were eagerly awaited by planetary scientists. The default expectation was that Europa would be a cratered, mountainous globe, much like Earth's moon, but the Pioneer pictures revealed a much stranger place, unlike anything astronomers had seen before.

Instead of a rugged landscape pocked with craters, Europa's surface is astonishingly smooth, encased in a shell of gleaming ice tinted with patches of ochre and yellow chemical deposits. Almost entirely lacking in mountains or other large vertical features, Europa is the smoothest body in the Solar System: if a billiard ball was scaled up to the same size, it would be lumpier than Europa. This absence of topographical relief – in particular the lack of large craters – indicates that Europa's surface is young in geological terms: like every other body in the Solar System, Europa must be struck periodically by asteroids and comets, but active geological processes ensure that its surface is being constantly renewed, erasing the evidence of all but the most recent impacts.

More surprises were in store when the twin Voyager probes swung past Jupiter in 1979, sending back more detailed views of the planet and its many moons. Europa may lack mountains and craters but its surface is far from featureless: the moon's icy shell is marked with a crazy paving of criss-cross fractures, resembling the ice floes of Earth's polar oceans, in which slabs of ice shift and grind against each other as currents move in the water beneath. Many scientists speculated that this resemblance was no accident, and that the reason for the youthful appearance of Europa's surface was that it was a crust of ice floating on an ocean of water.

Further evidence came from the Galileo probe, which went into orbit around Jupiter in 1995 and made several close flybys of Europa during its eight-year mission. Galileo confirmed the smooth, fractured appearance of the moon and found yet more signs of a subsurface ocean. The probe detected an interaction between Europa and Jupiter's powerful magnetic field that indicated the presence of an electrically conductive fluid within the moon. Since Europa's metallic core should long ago have cooled and solidified, the best candidate for this fluid is a layer of salty water beneath an outer shell of ice several kilometres thick. The current consensus is that the ocean itself could be around 100 kilometres deep, containing more than twice as much water as all of Earth's oceans combined.

Conditions in this underground sea must be strange indeed, with perpetual darkness, icy temperatures and crushing pressure, and yet it is considered to be one of the most promising places in which to search for extraterrestrial life. Liquid water is one of the three main requirements for living things, and as long as there are also sources of energy and nutrients – perhaps from geothermal activity on the sea floor – then the other requirements would also be in place. In the deep oceans of Earth, microbial ecosystems thrive around hydrothermal vents, where superheated, mineral-laden water emerges from cracks in the sea bed. Perhaps similar sites could also host life in the oceans of Europa? Finding out will be tricky – reaching the ocean itself will require not only a voyage across millions of kilometres of interplanetary space but then boring down through several kilometres of rock-hard ice. Nevertheless, scientists are already starting to think about how this might be achieved, and perhaps one day robot submersibles might cruise the dark ocean of Europa hunting for alien microbes.

Signs of life might be easier to detect on another ice-bound moon, this time orbiting the ringed planet Saturn. Tiny Enceladus is only around 500 kilometres across – about the size of England – but its small size belies its significance as another possible environment for extraterrestrial life. Enceladus is dazzlingly white – as white as freshly fallen snow – appropriately enough since, like Europa, its crust consists of frozen water. NASA's Cassini spacecraft has made several flybys of this moon, revealing a series of long parallel fractures in its southern hemisphere, which mission scientists have dubbed the 'Tiger Stripes'. More remarkable than the Tiger Stripes' appearance is what emerges from them: Cassini has imaged vast plumes of water vapour spraying from the cracks and reaching hundreds of kilometres into space, like giant versions of geysers here on Earth. The vapour freezes into tiny ice crystals, some of which fall back to the surface as 'snow', while others escape from the moon's gravity entirely and go on to form part of Saturn's extensive ring system.

The most likely source of the plumes is a reservoir of liquid water beneath Enceladus' crust of ice – perhaps even an extensive sea up to 10 kilometres deep. Like the ocean of Europa, such a body of water is a prime candidate for hosting simple microbial life and here at least the water is readily accessible since the moon is conveniently spraying it into space. Alas, Cassini's instruments were not designed to search for life, but the probe has flown through the plumes and directly sampled the water, finding evidence for salt and traces of organic molecules such as methane. Future missions will doubtless target the Tiger Stripes and their watery plumes as a window on the hidden liquid world beneath Enceladus' ice.

The moons Europa and Enceladus are currently the best bets for finding substantial bodies of liquid water elsewhere in the Solar System, but they are not the only candidates. Evidence is mounting that other ice-rich moons, such as Jupiter's Ganymede and Callisto and Saturn's Titan, probably also host subsurface oceans, and even the dwarf planets Pluto and Ceres are potential members of this club. If so, Earth's warm, open seas would actually be the anomaly and the vast majority of the Solar System's liquid water – and its potential habitats for life – would lie beneath the crusts of these frozen, outer worlds.

Beyond the Solar System, the potential for watery oceans is even more promising. Of the hundreds of planets discovered around distant stars, so far none are exactly Earth-like, but it seems overwhelmingly likely that our galaxy contains many other planets similar to our own, with seas, lakes and rivers – and perhaps even life. However, the huge haul of confirmed extra-solar planets has already thrown up some big surprises, and many of these worlds have properties unlike anything seen in our Solar System. This has set astronomers thinking about the range of possible planets that might be out there and how slight changes to the conditions of familiar-seeming worlds might turn them into altogether more exotic places. One of these hypothetical scenarios is that of the Ocean Planet or 'waterworld'.

We know of many objects in the Solar System that contain a very high fraction of water ice, including, as we've seen, moons of Jupiter and Saturn and dwarf planets such as Pluto and Ceres. We also know from studying the planetary systems of other stars that planets can sometimes migrate inwards, moving from cold, distant orbits to the warmer regions closer to their parent star. If a large icy body was to move closer to its star in this way, then it is possible that the ice would melt, entirely submerging the surface beneath a world-wide ocean of liquid water. Such an ocean could be very different from those that we're used to here on Earth, with depths of hundreds of kilometres (the deepest point in Earth's oceans is almost 11 kilometres, while the average depth is around 4 kilometres). The pressures at the bottom of these oceans would be immense, and might even crush the water into a strange form of ice that could exist at much higher temperatures than we are used to. Unsurprisingly, the atmosphere of a waterworld is likely to be saturated with water vapour, producing a strong greenhouse effect, and in extreme cases it could become 'supercritical', with so much water in the air that there would be no clear boundary between where the ocean stopped and the atmosphere began. On a moon-sized waterworld, the lower gravity might also allow enormous waves to form – a surfers' paradise, except for the lack of beaches on which they could break. Could these watery planets be hospitable to life? Perhaps: like the subsurface oceans of Europa, Enceladus and the other icy moons of the Solar System, it all depends on the presence of the right kinds of minerals and nutrients in the water. Even if waterworlds do exist, it will be some time before we are in a position to investigate them in such detail.

Oceans of an even stranger kind may be found on extrasolar planets orbiting extremely close to their parent star. These 'lava planets' would be rocky worlds in orbits much smaller than that of Mercury, the closest planet to the Sun in the Solar System. Here, gravitational forces would keep the planet 'tidally locked' to its star, with one face permanently bathed in intense heat and light and the other facing out into the darkness of space. Temperatures on the light 'day' side could be high enough to melt rock and this entire hemisphere would therefore consist of a deep ocean of liquid magma. Rivers of magma might flow around the planet to the colder 'night' side, which might also receive rainstorms of molten rock. Nothing like a lava planet currently exists in our own Solar System, but worlds of the right mass and at the right distance from their sun have already been discovered orbiting several other stars. Whether they really do possess lava oceans remains to be seen.

However, we don't need to travel to other stars to discover strange seas made of exotic substances – there is at least one example right here in our own Solar System. When NASA's Cassini spacecraft arrived at Saturn in 2004, one of the mission's primary goals was to investigate the planet's giant moon Titan. Larger than Mercury and almost as big as Mars, Titan is unique among the moons of the Solar System in that it has a dense atmosphere which, like that of the Earth, consists mostly of nitrogen gas. Saturn is almost ten times further from the Sun than the Earth is, and so the ringed planet and its moons receive only around one-hundredth of the solar heat and light that our own planet enjoys. With its nitrogen-rich atmosphere, Titan has been described as 'like Earth in a deep freeze' and scientists suspected that this chilly moon might be preserved in a primitive state, similar to the early Earth of billions of years ago.

When the Pioneer and Voyager probes flew past Saturn in the late 1970s and early 1980s, they returned images of Titan that were almost unbearably tantalizing. As well as nitrogen, the moon's atmosphere contains significant amounts of hydrocarbons, such as methane and ethane, that react with sunlight to create an opaque, orange smog similar to the haze of pollution that forms over cities here on Earth on a sunny day. Like the clouds of Venus, this haze completely screens the surface of Titan from view and the images of the moon showed only a fuzzy orange globe. Whatever the secrets of Titan, they would have to wait a quarter of a century for the arrival of Cassini, a bus-sized spacecraft specially equipped with infrared cameras and a radar system designed to penetrate the haze.

Cassini also carried a hitchhiker: the European Space Agency's Huygens probe, which was designed to parachute down through the atmosphere of Titan and land on the surface. In January 2005, Huygens made its descent, and as it dropped through the clouds, details of Titan's mysterious surface began to emerge for the first time. The first pictures from the probe, taken while it was still dozens of kilometres above the ground, revealed a vista of pale highlands interspersed with dark, smooth plains. But, as the images continued to arrive, the Titanian landscape revealed a surprise: between the hills were branching networks of dark channels, which seemed to wind their way down towards the plains just like river valleys here on Earth.

When Huygens finally touched down two and a half hours after first entering the atmosphere, it came to rest on one of these plains and sent back a solitary image of its surroundings. This final dispatch is the only picture ever taken from the surface of a moon other than our own, and it shows a flat, eerie landscape dotted with round pebbles up to several centimetres across, all beneath a misty orange sky. Like the winding channels seen from the air, these pebbles provoked a feverish reaction from the mission scientists. Rounded pebbles are common enough here on Earth, occurring on beaches and in stream beds all over the planet, but they had never been seen anywhere else in the Solar System. Pebbles are round because they have been smoothed, and this smoothing requires them to be rolled around by a flowing liquid.

A picture was emerging of a dynamic world with striking similarities to our own: rain falls from the dense orange clouds, courses through the streams and rivers of the mountains, and flows out on to the dark flood plains, where it deposits a cargo of silt and pebbles. There is just one problem: there is no liquid water at the surface of Titan. Although, like Europa and Enceladus, it is possible that the moon possesses a salty ocean deep beneath the ground, average surface temperatures are around -180 degrees Celsius so any water at the surface is frozen solid.

In fact, it turns out that on Titan water plays the role of rocks here on Earth: the rugged landscape is actually formed of ice, with the hills, rocks and pebbles being made of frozen water, while the dark sand is probably a sooty organic deposit that forms in the atmosphere and is washed out by the rain. But what is the liquid that rains from the clouds, carving and shaping the land below? Water is out of the question, but the temperature and atmospheric pressure on Titan are just right for another substance to exist in a liquid state – a substance that on Earth we know as a gas: methane. It is methane, probably mixed with its chemical relative ethane, that rains from the clouds of Titan, flows in its rivers and gushes across the plains – offering us a bizarre but instructive parallel to the atmospheric, hydrological and geological processes that shape our own world.

Huygens' mission is now complete and the probe will remain on the surface of Titan for centuries to come, perhaps becoming gradually buried in fresh deposits of icy silt and gravel as the long seasons of Titan take their course. But high above, Cassini has not been idle. As it orbits around Saturn its trajectory has periodically brought it close to Titan and each time the spacecraft has scanned the moon with its battery of cloud-piercing detectors, in the process making a further extraordinary discovery.

Cassini's radar system sends pulses of radio waves down to the surface, where they are reflected back into space, with the strength of the reflection determined by the roughness of the terrain. As expected, the mountainous highlands returned a strong radar signal, but as it scanned Titan's north pole the radar passed over large smooth regions that returned hardly any signal

at all. So smooth are these areas that the only plausible explanation is that they are lakes and seas of liquid methane and ethane.
Meanwhile, Cassini's cameras observed the moon using infrared
wavelengths, for which the Titanian haze is largely transparent.
These too revealed the presence of the seas – and were even able
to image sunlight glinting from their surfaces.

Apart from Earth, Titan is the only body in the Solar System that
is known to possess extensive, stable bodies of liquid open to the
sky. Several of the Titanian seas are comparable in size to the
Great Lakes of North America or the Caspian Sea of Eurasia.
They have intricate coastlines, fretted with fjords and cut by
wide river deltas and – like the 'seas' of our own Moon but with
far more justification – they are given the Latin title of *Mare* and
named after great sea monsters from Earth's mythology: Kraken
Mare from Norse folklore, Ligeia Mare after one of the Greek
Sirens, and Punga Mare after a water creature of Maori legend.

As yet, we can only imagine what an extraordinary scene these
methane seas must present, their oily surfaces glittering in weak
sunlight beneath an orange sky, and waves lapping against an
icy shore. But plans to return to Titan are already on the drawing
board and, although any future missions are still several decades
away, they promise to be truly extraordinary: exploring a sea
requires a boat and so engineers are busy designing craft that
could float on the Titanian waves – a 'space ship' in a far more
literal sense than we are used to.

The atmospheric chemistry of Titan holds up a mirror to the
earliest days of the Earth, giving us clues to the conditions that

prevailed here when our planet was young. But we now know that Titan's geology and weather also provide a strange, distorted reflection of Earth as it is today – a comparison that might even help us to gain a deeper understanding of our planet's current workings.

Earth is indeed a special planet, but we are not so different as we like to think. The seas that dominate our world and make it such a uniquely hospitable place also link us to seas in other places and other times – and each of these alien oceans has something profound to teach us.

Sunshine

Today, we can conjure light with the flick of an electric switch, and for most of human history before the nineteenth century it was fire, laboriously kindled and painstakingly tended, that gave humans some degree of control over the illumination of their surroundings. Yet, before they learned to harness electricity and fire, our ancestors were dependent on light from the sky – the Sun, Moon and stars – in order to see. One of our most basic senses – sight – was entirely reliant on astronomical objects for its functioning. It is the light of the Sun in particular that human vision has evolved to utilize. Even when our eyes are fully dark-adapted, with pupils dilated to admit every available scrap of light, our vision is operating at its very limits under the light of the stars. Moonlight can be much stronger but this borrowed light – actually just sunshine reflected from the Moon's rocky surface – is fickle: its timing and brightness varies throughout the month, making it an unreliable source of illumination, and if the night sky is cloudy often there will be no light at all to guide a nocturnal traveller.

Only the Sun is a reliable guarantor of vision, its rays penetrating all but the very thickest of clouds and ensuring that, during the hours of daylight at least, we're always able to see what we're doing. But the Sun's influence on us goes far beyond allowing us to see. Unlike the Moon and stars, the radiation that we receive from the Sun is so intense that it doesn't just light the Earth: it also warms it. This solar heating keeps our planet's average temperatures far above the bitterly cold levels typical of space, creating surface conditions on our planet that are conducive to the presence of liquid water – and therefore to life. The distribution of this heat over the Earth's surface also determines the structure of the planet's climate system and powers the short-term variations of the weather. The rotation of our planet causes the Sun to appear to circle the sky roughly once every 24 hours, rising in the east and setting in the west to give us the regular cycles of day and night on which our patterns of waking and sleeping and many other biological rhythms depend. The tilt of the Earth's axis means that, as it moves along its orbit, the maximum height of the Sun in the sky and the time it spends above the horizon vary throughout the year, giving us the familiar cycles of the seasons and determining the patterns of plant growth and animal behaviour that underpin our agriculture and farming (*see* Inconstant Moon, page 153).

The importance of the Sun was not lost on our ancestors. In many cultures it was worshipped as a god or served as a dazzling symbol of divine power, and for millennia its apparent daily and yearly motions across the sky were the principal means of telling the time, serving as the clock and calendar that are vital to the functioning of any complex society. Reflecting this importance, the Sun's motions have been ingrained in the architecture of ritual and religious sites down the ages and across the world, from the solstice alignments of Maeshowe and Stonehenge to the east–west orientation of Christian churches. On his deathbed, after a career spent capturing the effects of light, air, vapour and landscape on canvas, the painter J. M. W. Turner is said to have muttered 'The Sun is God.' Even in today's materialistic society, we still feel a profound connection to our local star: modern 'sun-seekers' and 'sun-worshippers' cross entire continents just to bathe in its rays, their resulting suntans testifying to their access to money and leisure. We might turn to science to justify these efforts – after all, exposure to sunlight is known to positively affect our mood and it is essential for the production of vitamin D – but it has also taught us to respect the Sun's power: its ultraviolet rays can harm our delicate skin, accelerating the signs of ageing and leading to cancer.

Go outside on a clear day and turn your face towards the Sun (make sure you close your eyes – staring at the Sun is never a good idea as it can damage your eyesight). You'll feel a sensation of warmth on your skin and see the red glow of sunlight filtering through the blood vessels of your eyelids. The infrared rays warming your face and the light hitting the retinas of your eyes are both forms of electromagnetic radiation. As they strike your body they transfer the energy that they carry to the atoms and molecules that make up your cells, and this is what is being picked up by the temperature receptors in your skin and the light receptors in your retinas. These cells in turn send electrical signals via your nerves to the brain, where they are translated into the familiar sensations of heat and brightness. Although we take such everyday experiences for granted, the biological machinery of the nervous system which allows us to feel the radiation of the Sun is astoundingly complex. But just as astonishing is the epic astrophysical journey that this radiation has taken in order to reach your body and give you the pleasant feeling of being outside on a sunny day. The welcome sunshine that lights our world and warms our skin has its origins 150 million kilometres away in an altogether less balmy environment: the searing nuclear furnace at the heart of our local star.

As we've already seen, the temperatures and pressures inside the stars are extreme, recreating those that existed in the first few minutes after the Big Bang itself. The Sun is no exception: deep in its core, with the full weight of 2 million trillion trillion kilograms of hydrogen and helium gas bearing down on top of them, atoms are stripped of their electrons, forming a plasma – a soup of positively charged atomic nuclei and negatively charged electrons, all of which are constantly jostling and colliding with each other. At temperatures of almost 16 million degrees Celsius, collisions between the particles are extremely violent: instead of simply rebounding from each other as atoms do under ordinary conditions like those here on Earth, in the centre of the Sun hydrogen nuclei are smashed together with such colossal force that the atoms merge, forming new, heavier atomic nuclei of the element helium. As they fuse together, a tiny fraction of the mass of the two hydrogen nuclei is converted into energy in the form of electromagnetic radiation.

The mass of a hydrogen atom is incredibly small – around a 600-trillion-trillionth of a kilogram – and the amount of mass lost during the merging process is even tinier – around 0.7 per cent of the total. But this minuscule quantity of matter translates into a substantial amount of energy – a relationship which is described by Einstein's famous equation $E=mc^2$, where E is the energy produced, m is the amount of mass converted and c is the speed of light. It's c that explains why the tiny mass converted during the collision is so significant: light is very fast – the fastest thing in the universe – at around 300 million metres per second. Multiplied by itself to give the c^2 of Einstein's equation it makes for a very big number indeed – 90,000 trillion – and so the amount of energy liberated is considerable.

Around 600 billion kilograms of hydrogen are converted into helium inside the Sun every second – this is around four times as much hydrogen as exists in all the water molecules (H_2O) of Lake Windermere, England's largest body of fresh water. Of this mass, around 0.7 per cent is converted into energy, so the Sun is losing around 4.26 billion kilograms of mass per second and transforming it into 400 trillion trillion watts of electromagnetic radiation (an energy yield equivalent to exploding 100 billion megatons of TNT every second). It is strange to think that the Sun is effectively disappearing before our eyes like this, but that is exactly what is happening. Luckily, 4.26 billion kilograms is only a very tiny fraction of the Sun's total mass and our parent star is in no danger of vanishing away entirely. Indeed, it has been losing mass at this rate for around 4.5 billion years and will continue to do so for another 5 billion – and, even after all that time, the amount of mass converted into energy will hardly have dented its original supply.

Each fusion reaction results in the production of a photon – a unit of electromagnetic radiation. But these are not yet the familiar photons of sunlight that human eyes have evolved to detect. Instead, they are gamma ray photons – the most extreme form of electromagnetic radiation, each one carrying hundreds of thousands of times as much energy as a single photon of sunlight. They have a long journey ahead of them before they can leave the Sun and shine out into space – and it is the rigours of this journey that will transform them into the much less energetic photons that we associate with sunshine.

When photons move through empty space there is nothing to stop them, and they can travel in a straight line over very large distances. Some forms of matter, such as air or glass, also allow photons of certain energies to pass through them unimpeded, and we say that they are transparent. But for the gamma ray photons, newly born from fusion reactions in the heart of the Sun, their environment is anything but transparent. Resembling conditions in the early universe, the heart of the Sun is an opaque fog of charged particles – positively charged atomic nuclei and negatively charged electrons – and no photon can travel very far without colliding with one of them. When it does the photon will be absorbed by the particle and then, after an interval, spat out again in a different direction – but with a little less energy than it originally carried. The energy lost by the photon is retained by the particle, and it this that helps to maintain the immense temperatures in the Sun's core.

The process of collision, absorption and ejection occurs again and again as the photons ricochet randomly through the centre of the Sun. Technically, a new photon is being created each time this occurs, so what is making the journey from particle to particle is not an individual photon but a succession of photons, each of which is the new vehicle for the energy of the original, minus the energy donated to the particle. In the dense environment of the core the average distance between successive collisions is only about one centimetre, and, since the solar core is hundreds of thousands of kilometres across and the direction of travel is constantly being altered, it takes an extremely long time for photons to emerge from the core into the outer layers of the Sun. In fact, the average time taken for a photon to exit the core is around 170,000 years, and some photons can take many times longer than this. Photons therefore spend a very long time trapped inside the Sun itself – and it is extraordinary to think that much of the sunshine that warms and illuminates our world was actually born in fusion reactions that took place in the heart of the Sun 170,000 years ago, when modern humans had only just evolved on the plains of Africa. Like the bones of our earliest ancestors, photons of sunlight are very much fossils from the distant past.

The Sun has a diameter of around 1 million kilometres and so, having made its arduous journey to the edge of the core, each photon still has to travel through hundreds of thousands of kilometres of the Sun's gaseous outer layers before it can emerge into space. Conditions outside the core are less extreme, however, and here the photons are even able to hitch a ride. Great clumps of superheated gas, each one thousands of kilometres across, rise up from the depths of the Sun like bubbles in a pan of boiling water, or columns of air rising over warm ground on a hot day – a process known as convection. Here, when photons are absorbed by particles of gas they are carried aloft for large distances before being re-emitted – and so, even though they are still constantly colliding and being absorbed, the photons' outward journey gets a helping hand. The bulk convective motion of the gas transports them away from the core – a journey that takes around a week.

and sea. But the nitrogen–oxygen mix of our atmosphere is largely transparent to photons of visible light: and so energy that has spent thousands of years forcing its way out from the heart of the Sun, then eight minutes speeding across 150 million kilometres of space, zips through Earth's atmosphere in a split second and triggers the light receptors of our retinas. From its liberation in the nuclear fusion reactions of the Sun's core to its manifestation as a nerve signal in our brains, the solar energy has travelled 150 million kilometres, handed down along a vast chain of individual photons.

The term 'visible light' refers to the range of photon energies that can be detected by the human eye. In the seventeenth century, Isaac Newton carried out a famous experiment in which he used a glass prism to disperse a beam of sunlight into a rain-bow, or spectrum, of colours. The principle behind this demon-stration is that photons of different energies are deflected through different angles by the glass of the prism, allowing them to be separated out, and our eyes perceive these different photon energies as distinct colours of light. The rainbows that we see in nature are themselves a manifestation of the same phenomenon, although in this case the dispersing medium takes the form of millions of water droplets rather than a single glass prism. Reversing this process shows that our perception of 'white' light is simply the combination of all the different photon energies given out by the Sun.

This highlights a rather strange situation: our eyes have evolved to perceive the Sun's combined light as pure white, so why then do we usually think of the Sun as being golden or yellow? In fact, the Sun's radiation does peak at the boundary between the yellow and green parts of the light spectrum, so it emits more yellow and green photons than any other colour. But, when combined with the spread of red, orange, blue, indigo and violet photons that are also present in sunlight, our eyes are designed to see the mixture as white, not yellow. The reason is still not fully understood, but it may be related to the fact that the Sun is usually too bright for us to look at directly, except when it is low in the sky.

Earth's atmosphere is mostly transparent to visible light but not entirely: air molecules and tiny dust particles can deflect the photons as they pass through, scattering them in all directions. The effect is more pronounced for photons at the blue end of the spectrum – these are scattered more readily and this is why the daytime sky appears blue. When the Sun is high in the sky, its rays shine down through about 100 kilometres of air and only the blue photons are significantly scattered. But when it is nearer to the horizon, in the morning and evening, the Sun's rays enter the atmosphere obliquely and must pass through many hundreds or even thousands of kilometres of air before reaching our eyes. This attenuates its blinding brightness, making it less uncomfortable for us to glance at it, but at these times the large paths traversed by the sunlight through the air also mean that many green as well as blue photons will be scattered from the Sun's direct rays. With these colours removed, this may explain why the Sun appears more yellow than white at the times of day when we are better able to look at it. The situation is even more extreme at sunrise and sunset when the path of the Sun's rays through the atmosphere is longest of all: only the reddest photons are able to travel directly from the Sun to our eyes and this is why the rising and setting Sun looks red.

As visually oriented animals, highly dependent on our sense of sight, it's tempting for us to think of the Sun's visible light output as the most significant way in which it affects our lives here on Earth. But, arguably, by far the most important influence of the Sun comes from its infrared radiation rather than visible light. As described earlier, most of the infrared photons arriving at Earth are either absorbed by the atmosphere, by the ground or by the oceans – in each case warming them with the energy that they carry. It is the Sun's infrared radiation that keeps our planet within the temperature range suitable for liquid water – and therefore in a state conducive to supporting life. We have the good fortune to be at just the right distance from the Sun so that the intensity of infrared rays arriving at the Earth's surface is neither too high nor too low – and our planet is thus neither too hot, like Mercury and Venus, nor too cold, like Mars and the other outer planets. This region of space around the Sun is known as the Habitable Zone (*see* Alien Oceans, page 75), and scientists suspect that the equivalent regions around other stars are some of the most likely places to look for planets capable of supporting life.

But the Sun's infrared radiation does more than just keep us warm. By heating the Earth's land, air and sea, it is the power source for the great currents of air and water that flow around the planet, forming the backbone of the entire climate system and the fuel that drives the short-term variations of the weather.

In the Earth's tropics the noonday Sun is almost directly over-head. Shining straight down, its rays have the shortest possible journey through the atmosphere and they strike the surface full on, transferring a maximum amount of heat to the ground and ocean. By contrast, close to the poles the Sun is mostly rather low in the sky. Approaching the surface at a shallow angle, the solar rays must penetrate a much greater thickness of air, which scatters and attenuates them, and when they strike the sea or land their warming energy is dispersed over a wide area. The effect is complicated by the daily revolution of the Earth, so that only half of the planet is receiving heat from the Sun while the other half is radiating its heat away into the cold and darkness of space (*see* Spinning Around, page 265). The Earth's axial tilt also leads to seasonal variations in day length and the maxi-mum height of the Sun in the sky but, averaged over a year, the tropics receive a much greater share of the Sun's energy than the poles, and it is this 'differential heating' that determines the Earth's main climate zones, as warm air and water flow around the planet in an endless attempt to even out the temperature differences between night and day, equator and poles.

Winds, waves and ocean currents are all manifestations of the Sun's influence, drawing their power from its infrared radiation. By providing the energy to evaporate water from the oceans, the Sun is also the source of clouds and rain – and thus of the fresh water that is essential for life on land and for geological processes such as erosion and soil deposition that sculpt the continents. We will explore in a later chapter (Spinning Around) how the spin of the Earth helps to distribute the Sun's heat around the globe, shaping its climate and weather patterns. These ceaseless movements of the atmosphere and oceans – from the Gulf Stream to the trade winds, and from storms to rain, to great rivers – have determined the course of human evolution and history as much as they shape our daily lives. All are powered by our parent star.

SOLAR POWER

With climate change and global warming high on the agenda, the words 'solar power' conjure up visions of clean, renewable energy, free from waste products such as heat-trapping greenhouse gas. The Sun, it seems, is the shiny new solution to our age-old energy problems. But in fact, for the whole of our history almost all of the energy used by human beings has come from the Sun – and many of our most polluting and environmentally damaging energy sources could quite justifiably be described as 'solar powered'.

Our civilization would be impossible without the ability to harness energy and direct it to our will, whether this is to move people and goods, to create light and heat or, increasingly, to collect, process and distribute vast quantities of information. Much of our technology is ultimately a means of transforming matter from one configuration to another and this process invariably requires a supply of energy, whether it's cooking, transportation or industrial manufacturing. Physics tells us that energy can never be created or destroyed, simply stored or transmitted – although, as Einstein showed with his equation $E=mc^2$, matter and energy can also be converted into each other and, as we've seen, this is the way that energy is generated inside the stars and in the Sun itself.

Energy manifests itself in many ways in the world around us: as the bulk motion of objects; as the small random motions of individual particles (which we experience as heat); as electromagnetic radiation such as light, X-rays, radio waves and infrared radiation (which can also interact with particles of matter to produce heat); in the directed flow of charged particles such as electrons (which we call electricity); or pent up within the chemical bonds between atoms. To access this energy, we must harvest it from our environment and convert it into the most useful form for our purposes.

For most of the history of the human species, we relied on the energy released in our own muscles to change and affect the world around us. Later this was supplemented by the use of tools, to concentrate and fine-tune our efforts, and by harnessing the muscle power of domesticated animals. In all these cases, the energy released comes from the metabolic processes within living cells in which sugars and other organic molecules are broken down by exposure to oxygen, producing carbon dioxide, water and energy. The molecular fuel has entered our bodies in the form of food, which, as animals at the top of the food chain, we humans have obtained from devouring the bodies of other animals or from the leaves, stems and fruits of plants. But, whatever our immediate source of nourishment (whether plant or animal), the molecules that it consists of were originally manufactured inside the cells of plants – and they were formed using the electromagnetic energy of the Sun.

The means by which plants achieve this feat is called photosynthesis, a word derived from the Ancient Greek for 'light' and 'putting together'. As its name suggests, this astonishingly complex chemical manufacturing process uses sunlight as the energy source to bind molecules of water (from the soil) and carbon dioxide (from the air) together, producing molecules of glucose and oxygen. (The oxygen in our atmosphere is therefore a by-product of plants' constant quest to harvest energy from sunlight – so even the air we breathe is a product of the Sun.) Glucose is easily transported around the plant and it acts as a kind of molecular 'battery', in which the energy that was used to create it is stored in the chemical bonds between its constituent atoms. This energy is waiting to be released again when the glucose molecules are allowed to react with oxygen – in the process releasing carbon dioxide and water molecules back into the environment and removing the oxygen that had been released in the original photosynthesis reaction. The energy from the glucose can then be used for a whole range of metabolic processes within living cells, from growth and movement to the manufacture of many other carbon-based, energy-rich molecules essential for life.

Of course, these energy-rich molecules can easily be stolen from the plants by any herbivorous animals that decide to eat them, by carnivorous animals that eat the herbivores and by saprophytic organisms that consume the decaying remains of plants, herbivores and carnivores alike. Photosynthesis – the trapping of the energy of sunlight within glucose molecules – is therefore the ultimate power source for almost all forms of life on the surface of the Earth. Our bodies – and those of most other living things around us – are quite literally solar powered.

When our ancestors learned to use fire to generate heat and light, they were again simply releasing the energy of trapped sunlight, which had originally been captured and stored inside the wood and other plant materials that they were burning. The steam-driven technology that later drove the Industrial Revolution used a refinement of this process, using the heat released by burning plant-based fuel to boil water into steam that could be used to push pistons and turn axles. Once again, there is a direct chain through which energy from the Sun is converted into useful work: the electromagnetic energy of sunlight, locked up in organic molecules within the plants, is released as heat, then converted into the bulk motion of the expanding steam which in turn is transferred into the motion of the machinery and directed, via clever engineering, into whatever task the machine was designed to perform.

With fossil fuels such as coal, oil and natural gas, humans discovered a way to improve the efficiency of steam technology still further. All these substances are the remains of long-dead plants and animals in which the energy (and carbon) harnessed by ancient photosynthesis has been locked away beneath the ground for millions of years. Fossil fuels have been compressed and concentrated by geological processes and they are therefore very rich sources of energy: they enabled a leap in power that helped to drive the Industrial Revolution to ever-greater heights of efficiency. Later, we also learned to convert their concentrated energy into a different, more versatile form, using steam to drive turbines and produce electrical currents that can be transmitted along conducting wires for hundreds of kilometres, then converted into heat, light and mechanical work or used to power a myriad of electronic devices wherever they might be needed.

Fossil fuels have proved to be an easy source of concentrated energy – all we have to do is dig them out of the ground and burn them and the energy of ancient sunlight, which once shone down on prehistoric forests and seas, is ours to use at will. The trouble is that there is only a finite supply of these miraculous fuels and, although they were laid down in slow, steady processes lasting millions of years, we have already raced through a large proportion of the available deposits in just a few centuries. Our industrial civilization is founded on an act of extraordinary profligacy, as we have proceeded to funnel the concentrated power of millions of years of photosynthesis into factories, power stations and vehicles over the brief span of modern history.

Aside from this extravagant use of ancient sunlight, the burning of fossil fuels has also had an unintended side effect. When we metabolize glucose molecules inside our cells, they are broken back down into water and carbon dioxide, and these molecules are expelled into the atmosphere when we exhale. Because there is a balance between the removal of carbon dioxide from the air via photosynthesis and its return to the air via respiration, overall levels of carbon dioxide in the atmosphere are not affected by this process. But, when we burn fossil fuels, as well as releasing the stored energy of ancient sunlight we are also releasing ancient carbon, converting it back into carbon dioxide and returning it to the atmosphere after an absence of millions of years. Since the Industrial Revolution began in the eighteenth century, the amount of carbon dioxide in the atmosphere has increased by around 43 per cent, and the heat-trapping properties of this greenhouse gas have led to a corresponding rise in global temperatures, which the majority of scientists fear will have profound environmental consequences.

In essence, then, the burning of fossil fuels is another form of solar power, just like the metabolic conversion of glucose that powers our cells. But coal, oil and gas are dwindling resources and they come with a damaging downside in the form of carbon dioxide emissions. Why risk altering the Earth's climate by resurrecting ancient photons of sunlight, when the Sun is still pouring its light over our planet every day? If we could capture and harness this solar radiation directly, then perhaps we could satisfy our hunger for energy without harming our environment.

Commercial solar power stations aim to realize this environmentally friendly dream by converting sunlight into electricity in one of two ways. Conventional solar panels make use of the 'photovoltaic effect' in which photons of light strike a cell made of a semiconductor material such as silicon, imparting their energy to electrons within its atoms and directly generating an electrical current. As well as involving no emissions of carbon dioxide, this process sidesteps the long and rather wasteful chain by which energy is converted from sunlight to useful work in traditional fossil fuel power stations. At each link in the chain, from photosynthesis to the burning of fuel, to the expansion of steam and the spinning of a turbine to generate electricity, a fraction of the energy is unavoidably lost, making the process wasteful as well as polluting.

In principle, the best solar cells can convert around 40 per cent of the energy of sunlight into useable electricity, although currently most commercially available cells have efficiencies closer to 15 per cent. By contrast, photosynthesis in plants is generally much less effective, with around 0.1–2 per cent of sunlight converted into useable energy – although some highly efficient plants such as sugarcane manage a more respectable 8 per cent. Any sunlight-harvesting scheme that relies on plants has already missed over 90 per cent of the available energy from the outset.

Another form of environmentally friendly solar power comes from solar concentrators. These use a less direct way of converting sunshine into electricity than solar panels: instead, mirrors or lenses focus sunlight from a large collecting area into a concentrated beam, which then heats a receptor material such as water, converting it into steam to drive turbines and generate electrical power in much the same way that a coal- or gas-fired power station would do. But, once again, no additional carbon dioxide is released in the process and so atmospheric levels of CO_2 are unaffected. Rather than converting water into steam, some experimental variants of solar concentrator technology use the focused heat to drive a chemical reaction in the hope of converting an even higher fraction of the Sun's energy directly into electricity.

However, plants and photosynthesis have not been entirely abandoned as tools for harnessing sunlight to meet human demands, as shown by the rising use of biofuels such as palm oil. Here, the solar energy is stored by the plant in the chemical structure of the oil molecule and is released once more when we burn the oil to generate heat, which can then be used to power an engine in a vehicle, or to drive a turbine and produce electricity for the grid. Unlike fossil fuels, in principle biofuels are carbon neutral: during photosynthesis the crop plant sucks carbon dioxide out of the air as a raw material for the production of the oil, so when we burn the oil we are simply releasing the same amount of carbon dioxide back into the atmosphere. But critics point out that biofuels can still lead to a net increase in global CO_2 levels, if the land they are grown on has been cleared of dense, natural forest in which large amounts of carbon was locked away in the trees themselves.

Encouragingly, 'clean' methods of harvesting solar power are becoming more and more widespread, and are beginning to make up a significant fraction of the world's total energy-generating budget. But direct solar power is only really viable in places where it is sunny for significant periods of time; at high latitudes, where winters are long and dark, or in temperate regions where skies are often cloudy, other methods of generating electricity are needed in order to supplement the rays of the Sun.

Luckily, these gloomier parts of the planet are blessed with copious storms, wind and rain – and the energy of the moving air and water can also be harnessed. Of course, this is nothing new: windmills and watermills have been used since ancient times to drive machinery. Modern wind turbines and wave energy converters now feed electricity straight into the power grid without adding to the carbon dioxide content of the atmosphere but, once again, the energy of the wind and waves is ultimately derived from the Sun. As we've seen, the Sun's infrared radiation warms the Earth's equator more than the poles, and this differential heating underlies both the long-term climate structure of the planet and the day-to-day variations of the weather, as currents of air flow around the globe in a never-ending attempt to equalize its temperature. The energy in the wind, and in the waves that it raises out at sea, is the energy of solar radiation. Wind and wave generators are yet another form of solar power.

There is another watery power source that also derives its energy from the Sun. In hydroelectric power stations, the downhill flow of water is used to drive turbines and thus generate electricity. But in order for this to work the water has to get to the top of the hill. In most cases, it arrives there as rain, falling from clouds that have condensed from water vapour high in the atmosphere. But how does the vapour get up into the air in the first place? It evaporates from the oceans as they are heated by the Sun's rays, rising into the atmosphere where it is then carried over the land by air currents. It is solar energy that raises the water, and it is this energy that is transformed into electricity in hydroelectric power stations as the water attempts to return to the sea.

Not all of our energy comes from the Sun, however. Tidal power, nuclear fission and geothermal power are not reliant on the rays of our parent star but, as we shall see in later chapters, they too utilize energy that has its ultimate origins in the depths of space.

There is a final energy source that – if we can ever get it to work – holds out the possibility of virtually limitless power without polluting the atmosphere with carbon dioxide, and here too the Sun has an important contribution to make. Nuclear fusion reactors mimic the processes that occur in the heart of the Sun itself, fusing hydrogen nuclei together to produce helium and, in the process, converting a tiny amount of mass into a substantial amount of energy in the form of radiation. As in other types of power station, the energy is used to heat water into steam and drive turbines to generate electricity. Currently, experimental reactors are able to achieve fusion, but the energy given out rarely exceeds the energy required to heat the hydrogen fuel and keep it contained for long enough for the fusion reactions occur. However, a commercially viable fusion reactor remains an attractive prospect: hydrogen fuel is in plentiful supply in the form of seawater, and nuclear fusion would largely avoid the problems of radioactive waste that plague our current nuclear fission reactors.

Nuclear fusion might liberate us from our dependence on the Sun's energy for the first time in human history and yet, even if it does, we will still owe it all to our parent star. Studies of fusion inside the Sun are playing a crucial role in enabling scientists and engineers to reproduce the process here on Earth. If they succeed, each fusion reactor will be like a miniature, artificial Sun – and so, in a sense, we will have learned to power our world not just from one star, but from many.

Electromagnetic radiation isn't the only thing given off by the Sun. As well as shining its photons into space, the Sun is also blasting out a torrent of charged particles, which streams away through the Solar System in all directions. This outflow of electrons and atomic nuclei – the debris of atoms wrenched apart in the inferno of the Sun's upper layers – is known as the 'solar wind' and, rather like winds here on Earth, its behaviour can be unpredictable. Around a billion kilograms of material is blasted from the Sun every second in this way, although the Sun is so massive that over its entire 4.5-billion-year career to date it has only lost around 0.001 per cent of its material to the solar wind. However, as we shall see in a later chapter (Written in the Stars), when the Sun reaches the end of its life the solar wind will play a significant role in its demise – and the demise of the rest of the Solar System.

The particles of the solar wind surge outwards at speeds of several hundred kilometres per second, passing the planets and filling a vast, bubble-like region known as the heliosphere, which extends well beyond the orbit of Pluto. Indeed, some scientists consider this outflow to be part of the Sun itself, so that the heliosphere could be thought of simply as the Sun's outer atmosphere – with the Earth and all of the other planets orbiting inside it. This may be stretching our definition of the Sun slightly too far but it is certainly true that the planets are continually exposed to the solar wind, and its buffeting has a considerable effect on all of them, including the Earth.

The solar wind is always present but it is anything but constant. Variations of temperature and magnetic field strength in the Sun's upper layers can affect the speed and density of the particles being launched into space and, since the Sun rotates on its axis roughly every 30 days, the Earth is exposed to a continually changing onslaught of gusts, squalls and flurries. Continuing the analogy with the fickle winds that blow in planetary atmospheres, the unpredictable fluctuations in the solar wind are known as 'space weather' and, like weather here on Earth, increasing amounts of effort are being spent on trying to make accurate forecasts of its behaviour.

The reason for all of this interest is that in recent years scientists and engineers have become increasingly aware that extreme incidences of space weather could have serious implications for much of the technology on which modern society depends. In particular, regions of the Sun's upper atmosphere occasionally erupt with enormous outbursts of material known as coronal mass ejections, which blast into space to form 'solar storms' many times more powerful than the background solar wind – and these pose a threat to technological infrastructure both on the ground and up in space.

It is somewhat paradoxical that, after millennia of being blithely ignorant of the very existence of space weather, the technological advances that have enabled us to learn about it have also made us highly vulnerable to its effects. The problem is largely down to the fact that both the ordinary solar wind and the more violent solar storms are made up of electrically charged particles travelling at very high speeds. This gives them similar properties to the nuclear radiation given off by radioactive substances such as uranium and plutonium, and they can cause similar problems for both living cells and electronic equipment.

Luckily, the Earth has two highly effective shields that protect it from the worst effects of the solar wind and more extreme space weather events. The first is our planet's geomagnetic field, which is generated deep in the molten core of the Earth but which extends for thousands of kilometres out into space, encasing us in a kind of magnetic bubble (*see* Spinning Around, page 265). The electrically charged particles of the solar wind find it difficult to cross this barrier and for the most part they are deflected around it, streaming past the Earth as if it was a rock in the middle of a stream. The only real chinks in the Earth's geomagnetic shield occur over the north and south poles of the planet, where the magnetic field lines dive back into the ground. Here the charged particles can find their way closer to the surface, and this is where our second protective barrier comes into play: the atmosphere.

High above the ground, usually at altitudes of between 90 and 150 kilometres, the charged particles from the Sun collide with atoms and molecules of air. In the process much of their energy is transferred to the atoms and is then released as photons of light – a similar process to that taking place in a sodium street lamp or a neon sign, in which atoms of gas are energized by an electric current. From the ground, the effects can appear spectacular, as great swirls and curtains of glowing gas shimmer overhead: these are the aurora borealis and aurora australis, the Northern and Southern Lights. Different gases in the atmosphere glow with their own characteristic colours – green and red for oxygen, blue, red and purple for nitrogen – and the combinations can produce beguiling displays that range from subtle delicacy to impressive grandeur. Though normally only visible from high latitudes, close to the poles, a powerful solar storm can expand the auroral zone, allowing the lightshow to be seen from latitudes nearer to the equator.

The aurorae are harmless and very beautiful, but they are also a demonstration of the Sun's pervasive influence – and a reminder of the protection afforded to our planet by its atmosphere. For satellites and astronauts orbiting above the Earth's insulating layer of air, the sight of the aurorae shimmering far below is less reassuring. Although still within the protective bubble of the Earth's magnetic field, people and objects outside the Earth's atmosphere are exposed to significantly higher levels of particle radiation than those on the ground. As well as the solar wind driving outwards from the Sun, space is also full of speeding particles of a more exotic kind. Like the particles of the solar wind these 'cosmic rays' are mostly composed of fragments of atomic nuclei but they are travelling at much higher speeds and they come from outside the Solar System. To date, the precise origins of cosmic ray particles remain a mystery, although their extreme speeds suggest that they are debris from some of the most violent events in the universe – including supernova explosions and the brutal environment at the centre of distant galaxies, where gas is funnelled into the mouths of supermassive black holes.

The energy carried by these subatomic projectiles can be huge – up to 40 million times the energies achieved in the particle accelerators of the Large Hadron Collider (LHC) at CERN – but fortunately for us the majority of them are absorbed by the Earth's atmosphere. When the LHC was first switched on in 2008 a handful of critics feared that the high-energy particle collisions produced in its 27-kilometre-long tunnel might trigger extreme types of physics that could destroy the planet. It was left to more level-headed physicists to point out that such violent collisions happen every day as cosmic rays slam into the Earth's upper atmosphere – indeed, they have been occurring for the last 4.5 billion years without seeming to bring about the end of the world.

However, for astronauts and spacecraft in orbit around the Earth, the cosmic rays and stray particles of the solar wind do pose a problem. On striking living tissue they can damage the DNA molecules at the heart of each cell, disrupting a host of biological functions and increasing the risk of cell death or cancerous growth. They can also cause havoc with electronic equipment, setting up dangerous currents in their circuits and damaging delicate components. Electronics can be 'hardened' against the effects of this radiation by careful (and costly) design, while the inhabited parts of piloted spacecraft can be insulated to some extent by thick (and also costly) shielding but, in the end, enhanced radiation levels are simply a fact of life beyond the Earth's atmosphere. After a six-month stint on the International Space Station (ISS), astronauts will have received a radiation dose around five times greater than their friends and families down on the ground, and space agencies devote a great deal of time and effort both to keeping this exposure to a minimum and to mitigating its potential effects.

At least the ISS is orbiting within the protective boundary of the Earth's magnetic field. If humans wish to travel further afield, they will have to contend with the harsh radiation conditions of interplanetary space, where they will be exposed to the full force of the solar wind. Robotic space probes to the planets and other Solar System bodies are carefully designed and shielded against its effects but for human explorers the need for protection is even more urgent. To date, the only people to have ventured beyond the Earth's magnetic sanctuary are the astronauts of the Apollo lunar exploration programme – and, even so, their time outside the protected zone was limited to just a few days. Although they all survived their trips, the radiation was not without its effects: Apollo astronauts reported seeing mysterious flashes of light, which were later explained as high-energy cosmic ray particles streaking through the fluid inside their eyeballs.

Radiation exposure remains one of the greatest obstacles to the more ambitious piloted missions to the Moon, asteroids and Mars that are being planned over the next few decades – missions that could last for many months, or even years. Unlike the Earth though, Mars, the Moon and many of the other proposed targets for human exploration lack their own magnetic fields, so even on arrival at their destination astronauts would find no respite from the solar wind and cosmic rays. Ingenious solutions are being pursued to reduce the risks, including the use of the spacecraft's drinking water tanks as an additional element of shielding or even generating an artificial magnetic field to surround the craft – a real-life echo of the protective 'force fields' so familiar to us from science-fiction stories. But, whatever the final strategy used, it will only be a partial solution to the problem, and increased radiation levels will remain an inevitable consequence of any voyage into space.

The unpredictability of space weather poses an additional problem. During a powerful solar storm the density of particles ejected by the Sun can be greatly enhanced and so the danger from radiation also increases accordingly. Luckily, the Apollo missions did not coincide with any dangerous solar outbursts but it is estimated that if one had occurred while the astronauts were outside their spacecraft they could have received a fatal dose of radiation within hours. Even within the bounds of the Earth's magnetic field, a solar storm is not an event to be taken lightly. Satellites are often powered down and placed in a dormant 'safe mode' for the duration of the storm to avoid damage to their components, while astronauts on the ISS take shelter in the more heavily shielded parts of the station. A typical satellite can cost hundreds of millions of dollars and the financial implications of a multiple satellite failure, leading to the shutdown of communications, weather monitoring or SatNav systems, would be immense – not to mention the havoc caused by the loss of essential services. But the interval between the initial eruption of a coronal mass ejection and the radiation particles reaching the Earth can range from a few days to just a few hours – an uncomfortably narrow window within which to issue warnings and put contingency plans into action. Space agencies, governments and companies are keen to protect their investments and much effort is being devoted to improving our understanding of the physics of the Sun so that better predictions of its behaviour can be made.

Technology down on the ground is also increasingly vulnerable to the effects of a powerful solar storm. Users of shortwave and amateur radio systems are familiar with the interference caused by even quite mild space weather events, but a major solar storm can cause problems for some forms of radar and military early warning systems as well as ground-to-air communications for aircraft. One of the most dangerous terrestrial manifestations comes not from the storm itself but from its effect on the Earth's magnetic field: as the charged particles strike the field they push against it, distorting its shape and forcing it inwards, closer to the Earth and causing rapid changes in the strength and direction of the magnetic field at ground level.

As you might expect, the changing field causes fluctuations in the direction of compass needles but the consequences are much more far-reaching than this. In particular, as the magnetic field sweeps over objects made of metal it can set up powerful flows of electricity, in the same way that moving a bar magnet inside a coil of wire will generate a current. The effects are more pronounced over very long distances and they are therefore a particular problem for transcontinental oil and gas pipelines and the long-distance cables used by power grids and the telephone network. In pipelines, the induced currents can confuse the readings of flow-meters and increase the rate of corrosion in the pipes, causing both short- and long-term difficulties for engineers, but the most serious problems of all come from the currents produced in power cables. In extreme cases, these can cause the electricity grid to overload, tripping safety systems and overheating transformers across the network, leading to major power failures.

A salutary demonstration of the Sun's disruptive power came on 13 March 1989 when a solar storm struck the Earth's magnetic field, sending currents surging through the cables of the Hydro-Québec electricity transmission system in Canada. In just 90 seconds, safety relays across the system were triggered one after another like toppling dominoes, shutting down the entire network and resulting in a power blackout that lasted nine hours and affected a total of 6 million people. The repair bill for the infrastructure of the grid came to millions of dollars, with knock-on losses to the customers of the power company amounting to millions more. Electricity providers around the world have learned from this incident, installing additional safety systems and ensuring that more robust warning procedures are in place. However, such precautions can never be 100 per cent proof against failure and, worryingly, the outburst in 1989 was relatively modest compared with the most powerful solar storm ever recorded.

On 1 September 1859, two amateur English astronomers, Richard Carrington and Richard Hodgson, were studying an unusually large group of sunspots – darker, cooler regions of the Sun's photosphere often associated with violent solar activity. The men were working independently but, just before noon, both observed a strange phenomenon: a bright patch of light suddenly blossomed above the sunspot cluster, blazing intensely for around five minutes before fading away. This was a solar flare – an explosion of energy in the Sun's atmosphere – and it triggered a coronal mass ejection on an unprecedented scale. A vast cloud of charged particles was blasted from the Sun on a direct collision course with the Earth and, 17.5 hours later, this colossal solar storm struck our planet's magnetic field head on.

Intense auroral displays lit up the night sky over a huge area, appearing as close to the equator as Mexico, Cuba and the Sahara in the northern hemisphere and Queensland and New Guinea in the south. Over North America the aurora was reportedly bright enough to read by, while some observers mistook its glow for the approach of the dawn. The international telegraph system was also affected: sparks flew from telegraph poles while operators received electric shocks as currents surged along the cables and into their sets. Many parts of the network were overloaded while others spookily continued to operate on the induced currents, even when their normal electricity supplies had been disconnected. Scientists were quick to draw connections between the unusual events here on Earth and the intense solar flare observed just a few hours previously, giving them a new and disconcerting insight into the extent of the Sun's influence on our planet. Now known as the Carrington Event, this solar storm helped to galvanize the emerging field of solar physics, but if a similar solar storm was to occur today its effects could be far less benign.

In 1859, the telegraph system was the only widely used technology that was powered by electricity and so, apart from a few unfortunate telegraph operators, the world could marvel at the display without too much concern. But now we live in an electrical world, with almost every aspect of our daily lives dependent on power from the grid and on information transmitted around the globe via satellites. In the twenty-first century a Carrington-scale solar storm could cause a planet-wide failure of electricity networks, plunging our infrastructure back into the nineteenth century. Sourcing the components to repair thousands of electricity transformers and get the power systems up and running again would be a mammoth task, and billions of people could be without power for months – with a total cost to the global economy running into trillions of dollars. Replacing whole fleets of damaged and destroyed satellites would take even longer and cost billions more – with services taking many years to be fully restored.

As long as we have sufficient warning, action could be taken to avoid this worst-case scenario. Power grids could be temporarily shut down, electricity transformers isolated and orbiting satellites placed in safe mode until the storm passes. A few hours with only basic emergency services might be inconvenient but at least we could sit back and enjoy the spectacular auroral light show safe in the knowledge that things would soon be restored to normality. Several countries, including the United Kingdom, now include solar storms in their national risk registers and protocols for monitoring and reacting to dangerous solar outbursts are starting to be drawn up.

But time is of the essence: in 2014 NASA released information on a solar storm of similar magnitude to the Carrington Event, which had occurred on 23 July 2012. Luckily, the outburst was not directed towards the Earth and our planet escaped unscathed – this time. However, this incident highlights the fact that studying our cosmic environment is not just an exercise in idle curiosity. A better understanding of the detailed physics of stars will be essential for predicting the future behaviour of the Sun and allowing us to prepare for and mitigate the worst effects of the next Carrington Event when it occurs – as it inevitably will. In a world reliant on satellites and electronics, ignorance is no longer an option.

There is one piece of human technology that is no longer subject to the vagaries of the solar wind. On 25 August 2012, after a journey lasting 35 years, NASA's Voyager 1 spacecraft passed through the outer boundary of the heliosphere – where the solar wind finally falters – at a distance of 18 billion kilometres from the Sun, becoming the first ever object created by humans to enter interstellar space. Prior to this, the whole of human history and civilization – the whole of our biological evolution in fact – had taken place within the embrace of our parent star, but now we have sent a representative of our culture outside the Sun's sphere of influence. In the coming years, Voyager 1 will be joined by its sister craft Voyager 2, the twin Pioneer 10 and 11 probes and the New Horizons mission, heading unstoppably onwards after its encounter with Pluto in 2015.

As well as the day-to-day fluctuations of space weather, we have come to understand that the Sun's behaviour varies on a number of much longer timescales, and these changes can also have a profound effect on our planet. For centuries, astronomers have been fascinated by sunspots, dark patches that temporarily form in the Sun's photosphere and persist for days or weeks before fading away. In the sixteenth and seventeenth centuries, observations of their transitory nature provided an important piece of evidence that contradicted the traditional model of the universe, dating back to the Ancient Greek philosopher Aristotle, that heavenly bodies like the Sun were perfect and unchanging. We now understand that the sunspots' dark appearance is due to the fact that they are cooler than their surroundings, although 'cooler' is a relative term – their average temperatures are between 3,000 and 4,000 degrees Celsius, while the typical temperature of the photosphere is 5,500 degrees Celsius. They occur in regions where the Sun's own internal magnetic field has become distorted causing magnetic field lines to arch out of the photosphere like croquet hoops on a lawn. These magnetic loops suppress the upward convection of hot gas from the solar interior resulting in regions of reduced temperature, which can often be as large as the Earth itself. But the field lines are in constant motion, writhing and coiling within the Sun's outer layers, and so sunspots are ephemeral by nature.

In 1847, after observing the Sun for 17 successive years, the German astronomer Samuel Heinrich Schwabe noticed a periodic variation in the number of sunspots that could be seen from year to year. Intrigued by this, his Swiss colleague Rudolf Wolf compiled records of sunspot counts dating back more than two centuries to the time of Galileo, and a clear pattern began to emerge: roughly every 11 years the number of sunspots visible on the Sun's disc reaches a peak, declining to a minimum 5.5 years later before building up to another maximum and beginning the cycle again. The peak of this cycle is called the solar maximum.

After more than a century and a half of study, there are many details of the sunspot cycle that remain a mystery, but we do now understand its underlying cause. Unlike the Earth, the Sun is not a solid body and its gaseous outer layers rotate about its axis at different rates depending on latitude, ranging from around 25 days for a full revolution at the solar equator to just over 34 days close to the poles. The Sun's internal magnetic field is naturally oriented along the direction linking the poles, but as the Sun spins, its 'differential rotation' winds the field around the equator, in the process causing it to become increasingly tangled – like a ball of wool being toyed with by a playful kitten. The winding increases for several years until the strain becomes too much, at which point the magnetic field realigns itself and the process begins again.

As the magnetic field becomes ever more twisted and chaotic, parts of it loop out through the photosphere, giving rise to sunspots, and, when the stress in these loops becomes too great, the field lines can violently snap, releasing a burst of energy in the form of a solar flare. The frequency of solar flares and coronal mass ejections therefore increases along with the number of sunspots, and so solar storms and other episodes of extreme space weather are more likely to occur around the time of the solar maximum.

The Sun also brightens by around 0.7 per cent during solar maximum, causing a measurable variation in the Earth's climate – although this is much smaller than the effects of global warming due to our own greenhouse gas emissions. (It might seem odd that the Sun's total brightness increases at exactly the time when dark, cool sunspots are at their most common, but the reason is that the rest of the Sun brightens by an amount that more than compensates.)

Towards the end of the nineteenth century, the astronomers Gustav Spörer, at the New Berlin Observatory, and the husband and wife team of Walter and Annie Maunder, based at the Royal Observatory in Greenwich, looked again at the historical records of sunspot numbers and realized that not all sunspot cycles are created equal. In particular, they identified a period from 1645 to 1715 during which hardly any sunspots were observed at all, even at the 11-year peaks of each cycle. This was not because nobody had been looking for them: on the contrary, highly respected astronomers of the period, such as Giovanni Domenico Cassini at the Paris Observatory and Johannes Hevelius in the Polish city of Gdańsk, had carried out meticulous observations of the Sun. Instead, it seemed that sunspot numbers really were greatly suppressed during this period, with a typical 11-year cycle featuring just a few tens of sunspots, compared with the tens of thousands that appear per cycle today.

This 70-year dip is now known as the Maunder Minimum (another dip, from 1460 to around 1550, is named after Spörer) and, although scientists still do not fully understand what caused it, it seems that it too may have made its mark on our planet. The Maunder Minimum coincides with the middle section of the so-called 'Little Ice Age', a period during which Northern Europe suffered a series of bitterly cold winters. The term Little Ice Age is somewhat misleading – not every winter during this time was especially cold, and summers seem to have been no cooler than usual. But it seems that the likelihood of harsh winters was certainly higher than it is now, and the Sun's unusually subdued behaviour may explain why. Computer models suggest that the suppression of sunspots would have been accompanied by enhanced emission of ultraviolet (UV) radiation. These additional UV photons would have been absorbed by the upper layers of the Earth's atmosphere, causing them to expand, and in turn this could have diverted the path of the jet stream, the high-altitude 'river' of air that winds around the planet, making it easier for cold, Arctic air to flow down over Northern Europe during the winter months.

The connection is controversial and is still the subject of active debate, but if it turns out that the Maunder Minimum really was implicated in Europe's run of cold winters, then the Sun's odd behaviour has left us with a cultural legacy. The Little Ice Age is famous for the freezing of the River Thames, allowing Londoners to hold 'Frost Fairs' on its icy surface. This freezing was largely due to the fact that the Thames was a more lethargic river prior to the nineteenth century, when the construction of embankments narrowed its course and the replacement of the original London Bridge by a new structure with wider arches allowed a greater flow of water, but there's no doubt that the unusually cold temperatures also played a part in creating the right conditions for the winter festivities. Some commentators have suggested that a profusion of striking wintery scenes in the art of the period – including famous examples by the Netherlandish painters Pieter Breughel the Elder and his son – might also be a response to the prevalence of severe winters. And there may even be a musical connection: the exquisite tone of violins crafted by the instrument maker Antonio Stradivari during the Little Ice Age has been linked to the quality of their wood, and perhaps this too was shaped by the cold conditions.

Whatever the truth of these suggestions, we cannot escape the fact that the Sun is the most significant object in our skies, exerting its influence on planet Earth in a huge variety of ways. Rather than resenting its fickle nature, we should perhaps be grateful that its unpredictability is limited in scale. Recent studies have shown that some apparently sun-like stars occasionally produce 'superflares' – explosions of energy up to a million times more powerful than the solar flare that triggered the Carrington Event in 1859. If such a superflare were to occur on the Sun, its effects on Earth would be truly catastrophic. The Sun would temporarily brighten by a considerable factor, perhaps enough to trigger the melting of ice sheets in the Arctic and Antarctic. When the associated solar storm struck the planet a few hours later, it would overwhelm our magnetic shield and bombard the upper atmosphere, destroying the ozone layer and delivering a potentially fatal dose of radiation to anyone unlucky enough to find themselves in an aircraft at cruising altitude. Luckily, there is no evidence that the Sun has ever suffered a superflare at any time during the last 4.5 billion years and it seems likely that, despite their superficial resemblance, the handful of stars that have been seen to exhibit superflare behaviour actually differ from the Sun in some as yet undetermined way.

Still, these extreme events serve to remind us that we should never take our comfortable lives in this quiet corner of the universe for granted. All taken with all, the Sun is an uncommonly well-behaved star and its steady output of energy has allowed life on our planet to flourish for billions of years. This state of affairs will continue into the future for a considerable time, but eventually even the Sun will die – and the Earth will inevitably share its fate. But that is a story for a later chapter (see Written in the Stars, page 308).

Inconstant Moon

A part from the Sun, the most conspicuous object in our skies is Earth's nearest astronomical neighbour – and only natural satellite – the Moon. The Moon is the only place in the universe aside from the Earth on which human beings have set foot: between 1969 and 1972 twelve people walked on its ashen surface and their footprints remain to this day, undisturbed by wind or rain in the Moon's airless environment.

Shining by reflecting sunlight rather than emitting light, the Moon is often visible by day – the only astronomical object that can make this claim – and it can outshine everything else in the night sky. But the Moon is not a reliable light source: the amount of illumination that it provides is constantly changing as it goes through its endless cycle of phases, waxing from New Moon to crescent, gibbous and Full Moon, and then waning back through the sequence to begin again at New Moon. The Moon's changeable appearance is due to the fact that, like the Earth, only half of its globe is lit up by the Sun at any one time and, as it orbits around us, the fraction of this lunar day-side that is visible from Earth constantly alters.

Although moonlight is really just borrowed sunshine, Earth's natural satellite has nevertheless played a major role in human history and culture. For ancient societies the Moon's regular cycle of phases, repeating on average every 29.5 days, was one of the main ways of keeping track of the passage of time. Even today, the Islamic calendar is based on the phases of the Moon, and many other cultures use a hybrid 'lunisolar' calendar alongside the western Gregorian calendar, which is based on the orbit of the Earth around the Sun.

To this day, the Moon's calendrical importance is reflected in the English language, since the word 'month' shares a common derivation with 'moon' – a linguistic reminder of a time when a month was simply one complete cycle of the lunar phases. The months of the western calendar were long ago padded out with extra days and no longer keep track of the lunar cycle. This is due to the inconvenient fact that 12 lunar cycles of 29.5 days is still several days short of a solar year, as defined by one complete orbit of the Earth around the Sun. The duration of the Moon's orbit around the Earth and the Earth's orbit around the Sun are both accidents of the way the Solar System formed, and there really is no reason why one of them should be a perfect multiple of the other. To stop the calendar months from drifting out of step with the year – and with the seasons – we have divorced them from their original lunar association.

As well as keeping track of time, the Moon has played a vital role in helping us to find our way around. Before streetlights were common in towns and cities, journeys by night were much easier and safer around the time of a Full Moon, when its illumination would allow travellers to see their way clearly. And this is not the only way in which the Moon has enabled us to navigate. In the seventeenth century, the Royal Observatory was founded in Greenwich specifically to make accurate measurements of the positions of the stars and the motion of the Moon, so that they could be used by ships to find their position at sea, when out of sight of land (*see* White Dielectric Material, page 39). This 'lunar distance method' of navigation was widely used by sailors well into the twentieth century, and even today it is still taught as a backup in case a ship's electronic navigation systems, such as the Global Positioning System, should fail.

But the Moon's influence on the Earth extends far beyond our own desire to tell the time and find our way around. At 3,474 kilometres across, the Moon is too small and its gravity too weak for it to have retained any atmosphere that it may once have possessed. Due to its smaller size, the Moon's interior cooled and solidified more rapidly than that of the Earth and it no longer generates a magnetic field to protect it from the solar wind. Its barren, airless surface is exposed to extreme temperature swings and the harsh radiation of space, and the Moon is a permanent reminder of our own good fortune in living on a planet that is wrapped in a blanket of air and protected by an electromagnetic shield. And yet, despite having just 27 per cent of the diameter and 1 per cent of the mass of the Earth, the Moon has helped to shape our own planet ever since it first formed. In fact, many scientists believe that without the Moon the Earth would have turned out very differently – and may not even have been habitable at all.

The Moon may be far smaller than the Earth but it is nevertheless a giant among the natural satellites of the Solar System, ranking fifth in both size and mass after Ganymede, Titan, Callisto and Io. These other top-ranking moons belong to the gas giant planets Jupiter and Saturn, and Earth is unusual among the rocky, inner planets of the Solar System in having such a large satellite. Mercury and Venus have no moons at all, and although Mars has two moons, they are both tiny, irregularly shaped objects just a few kilometres across. So unusual is the Earth's situation that it has sometimes been suggested that the Earth–Moon system should be regarded as a double planet with both objects orbiting each other. However, although it is true that the Earth and Moon both orbit a common centre of mass, this point still lies within the Earth itself, about 1,710 kilometres below the surface, and so we can legitimately claim that the Moon is orbiting around us: a true satellite of our planet.

Interestingly, this is not true of the dwarf planet Pluto and its largest moon Charon. Charon has almost 12 per cent of the mass of Pluto, and the centre of mass of the Pluto–Charon system lies well outside of Pluto itself: both objects are in orbit around this point and so technically they constitute a binary system rather than a parent body and its satellite. Pluto's four other known moons are much smaller than Charon and they all also orbit the centre of mass of the Pluto–Charon system rather than Pluto itself, making this one of the strangest orbital configurations in the Solar System.

There have been many theories about how the Earth came to acquire its unusually large satellite. In the nineteenth century, it was suggested that the newly formed Earth was spinning so rapidly that a huge chunk of rock was thrown out from the equator and into orbit; but for this to happen, the rate of spin required would be phenomenal, and far beyond any plausible scenario for the birth of our planet. Another suggestion was that the Moon was formed elsewhere in the Solar System and was only later captured by the Earth, but it is very hard to imagine how the Moon could have lost sufficient orbital energy to be trapped by Earth's gravity. Or perhaps the Earth and Moon formed from the same cloud of rocky debris, so the Moon has always been here. But in this scenario, how do we explain the subtle differences between the composition of the rocks that make up the Earth and the Moon?

Today, the favoured theory for the Moon's origin involves a catastrophic impact between the early Earth and 'Theia', a hypothetical planet similar in size to Mars. Named after the goddess who was the mother of the Moon in Greek mythology, Theia probably formed and orbited the Sun in the same orbit as the Earth, at a gravitational 'sweet spot' equidistant from both the Earth and the Sun, known as a Lagrangian point. About 4.5 billion years ago, as the planet-building process drew to a close, the periodic gravitational tugging of the other worlds of the Solar System began to exert a fatal influence on Earth and Theia, destabilizing their delicate orbital balance until the smaller Theia was dislodged and set on a collision course with its neighbour.

It seems that Theia must have struck the Earth a glancing blow, but the energy of the impact was still tremendous, sending a vast spray of molten debris into space. The metallic core of Theia sank into the Earth to merge with the core of our own planet, while trillions of tons of rock were vaporized and the surface of the Earth was reduced to an ocean of magma. Within a century or so the orbiting debris would have clumped together under the influence of gravity and the Moon was born, although at this early date it would have been ten times closer than it is now and it must have presented an astonishing sight – an angry globe of glowing magma, dominating the sky above the Earth's own molten landscape.

Extreme though it sounds, this 'giant impact hypothesis' now has broad support from the planetary science community, and ironically it may be the violence of the collision itself that was instrumental in creating the conditions under which life could later flourish on Earth. Although some details remain to be worked out, Theia's glancing impact fits with the orbital configuration of the Earth–Moon system and it also explains why the Moon, which was formed largely from the rocky outer layers of Theia, is depleted in some metallic elements relative to the Earth, which received the bulk of Theia's metallic core.

It is this influx of metals into the Earth's rocky mix that could be regarded as the Moon's first beneficial legacy to our planet. For much of its formation process, the Earth had grown by accreting vast numbers of much smaller objects such as asteroids and comets, but, although individual impacts would certainly have been impressive, by the late stages of this process the Earth would have been so large that its molten interior would have remained mostly undisturbed by the pounding of the surface. Under these conditions, heavier elements such as metals would tend to sink into the centre, leaving the surface layers of the Earth relatively devoid of these valuable materials. But the impact of Theia was on a different scale entirely: it would have sent shockwaves deep into the planet, churning up the interior as its metallic core sank into the centre to merge with that of the Earth. This may be the reason why our planet's crust contains so many useful deposits of metallic ores, which 4.5 billion years later proved to be essential raw materials for the development of our technological civilization.

This initial gift of metals was by no means the only legacy of the Moon's tumultuous birth. By blasting huge amounts of rock into space, the collision may have ensured that our planet's outer layers were thin enough for plate tectonics to establish itself once the Earth had cooled sufficiently for the surface to solidify. Plate tectonics causes the continental land masses to rearrange themselves over geological timescales, creating a slowly changing pattern of climates and coastlines that has shaped the evolution of life for billions of years. And by continually recycling rocks and driving volcanism, it also plays an ongoing role in the great geological, oceanic and atmospheric cycles that help to keep conditions on the surface of our planet within a range that is tolerable for life.

Venus, our nearest planetary neighbour, gives us an insight into how differently Earth might have turned out were it not for its early encounter with Theia. In some ways Venus is our twin planet, with a diameter at 95 per cent and mass 82 per cent of Earth's. Yet there are profound differences between the two worlds: Venus has surface temperatures of 450 degrees Celsius and the planet's crust shows no sign of mobile tectonic plates. These circumstances appear to be linked. Without plate tectonics to remove carbon dioxide from the atmosphere, the planet is locked into a relentless cycle of greenhouse warming. On Earth, plate tectonics relies on water from the oceans to lubricate the movement of the crust but on Venus the atmospheric greenhouse effect keeps surface temperatures far too high to permit liquid water to exist, and so the situation is perpetuated. Despite this planetary Catch-22 scenario, it is likely that, just like Earth, Venus started out with substantial oceans – so why did plate tectonics not establish itself on Venus as it did on Earth? Perhaps part of the answer lies in the absence of a Venusian moon. Venus has undoubtedly suffered many impact events during its 4.5 billion-year history but if it never underwent a collision on the same scale and of the same glancing nature as the Earth–Theia encounter, then this may have left it with a crust that is simply too thick to develop mobile plates. If so, the hellish conditions on Venus are a sobering reminder of our own good fortune.

It is also likely that the collision with Theia was the defining event that set the tilt of the Earth's axis. Because of this tilt our planet's northern and southern hemispheres receive changing amounts of heat and light during the course of Earth's yearly orbit, as first one hemisphere and then the other leans towards the Sun, and this is what creates the familiar cycle of the seasons. By determining how much solar energy is received at different latitudes at various times of year, the angle of tilt is also one of the defining factors behind the positions of the Earth's main climatic zones, from the tropics through the temperate latitudes and up to the Arctic and Antarctic regions.

Even today, the Moon continues to exert a profound influence on our planet's tilt. The physics of rotating objects means that they are prone to instabilities, so that the slightest disturbance will cause them to sway back and forth as they spin – a familiar concept to anyone who has ever watched a spinning top wobbling from side to side on a flat surface. A spinning planet is no exception: regular gravitational tugs from the other planets as they move around their orbits can cause its axis of rotation to sway and wobble chaotically from side to side over periods of just a few thousand years. However, as far as we can tell, the Earth's axial tilt has remained remarkably stable for hundreds of millions of years, varying by only a degree or so – much less than we would expect given the gravitational influences of our planetary neighbours. Something is protecting us from these interplanetary disturbances, acting to dampen the effects of their regular tugging, and it turns out that our protection comes from the gravity of the Moon.

The principle is similar to that of a tightrope walker who carries a long horizontal pole to help her balance on the narrow wire: if she sways too far to either side she will overbalance and topple off, but the pole acts as a stabilizer, dampening the effect of small swaying motions so that only a very strong sideways force would be enough to dislodge her. Like the tightrope walker, the Earth is prone to instabilities, but by exerting its gravitational influence from 400,000 kilometres away the Moon plays the same role as the horizontal pole, stabilizing our planet against its natural tendency to sway back and forth as it spins.

It's hard to say exactly how important this lunar damping system has been for the evolution of life on our planet but, without it, over periods of just a few thousand years, the Earth's axial tilt could have varied from 0 degrees, with no seasonal variations at all, to 90 degrees, with each hemisphere spending large portions of the year either in continuous daylight or permanent night. The effects on our climate would be immense, putting current fears about climate change and global warming into the shade. At any point on the globe, temperatures could swing from tropical to arctic in just a few centuries, while ocean currents and patterns of atmospheric circulation would be continually adjusting to the changing distribution of solar heat. This would make nonsense out of the notion of fixed climate zones, and ecological systems would barely get a chance to establish themselves in one location before the climatic conditions changed yet again. Under such circumstances, it might still be possible for simple microbial life to form and thrive but it is by no means certain that complex plants and animals would ever manage to gain a foothold.

Again, one of our neighbouring worlds provides a sobering demonstration of how the Earth might behave were it not for the restraining influence of the Moon. Mars currently has an axial tilt of about 25 degrees, giving it a climate that varies seasonally in similar ways to the Earth's, with winters and summers during which the planet's icecaps expand and contract. However, the Red Planet lacks a large moon and it's likely that these current rather modest seasonal variations are not a permanent state of affairs. Indeed, computer simulations indicate that over millions of years the planet's axial tilt might have ranged from as little as 0 degrees to as much as 60 degrees, leading to climatic conditions very different from those that it experiences today.

Even with the Moon's restraining influence, the Earth's axial tilt still varies, and, at present, the tilt is decreasing at around 47 arcseconds (sixtieths of a sixtieth of a degree) per century.

Currently, the tilt is 23.4 degrees, but over a period of about 41,000 years it changes cyclically, from as little as 22 degrees to as much as 24.5 degrees, with a corresponding effect on the severity of seasonal variations: less intense seasons when the tilt is small and more intense when it's large. While it works to damp down large-scale changes, the Moon itself causes the Earth's axis to 'nod' back and forth by a tiny amount – about 9 arcseconds every 18.6 years – in a process called 'nutation', which was first discovered by Astronomer Royal James Bradley in the eighteenth century (*see* And Yet It Moves, page 218). The variation of a few degrees can still have a significant effect on our climate and, along with changes in the shape of the Earth's orbit around the Sun, it is responsible for regularly plunging the Earth into the cold, glacial phases that we know as Ice Ages. If even a small change in our angle of rotation can have such dramatic consequences, it only goes to show how important the Moon has been in ensuring that conditions on Earth remain within the range suitable for life.

MOONSTRUCK

For 4.5 billion years the Moon has made its presence felt here on Earth, and its legacy is all around us in the shape of our planet's geology, oceans and climate. Since our species first evolved around 200,000 years ago, the Moon has also worked its wiles on us, exerting a profound influence on human culture. Everywhere we look, we can find evidence of the Moon's influence, from the divisions of our calendar to our art and literature – and even our language.

With its pale light and constantly changing appearance, the Moon has become an enduring symbol of both romance and fickleness. Poems and songs, from Sappho's 'Full Moon' to the 'Blue Moon' of Rodgers and Hart, testify to the Moon's amatory power, while Shakespeare's Juliet entreats Romeo to 'swear not by the moon, th' inconstant moon . . . lest that thy love prove likewise variable'.

The Moon's dominance of the night sky has also led to its association with dark forces and occult powers. Perhaps the best known of these associations is the myth of the werewolf, a human being who is transformed into a savage monster by the power of the Full Moon. Here, the lunar connection may originate in the popular belief that dogs and wolves howl at the Moon, although in the original European folktales the trigger for the werewolf's transformation was just as likely to be a magic potion or a pact with the Devil as any direct influence from the Moon itself.

For centuries, madness was also believed to fall under the influence of the Moon and the word 'lunatic' is a direct derivation from the Latin word for moon, *luna*. Even today, there are occasional claims from police forces and hospital staff that the rates of violent crime and erratic behaviour increase during the nights around a Full Moon, but detailed studies of large sets of crime statistics have failed to show a strong correlation. Perhaps this belief stems from folk memories of the period before widespread street lighting, when bright moonlight meant that people were more likely to be out and about at night. Coupled with the well-known phenomenon of 'confirmation bias', in which we give more significance to events that support our existing beliefs while discounting those that don't fit the expected or desired pattern, this might explain why we continue to link unruly behaviour with the Full Moon despite a lack of solid evidence.

It is certainly true that the Moon has an effect on animal behaviour. Some nocturnal birds such as eagle owls call and display when the Moon is full, while many nocturnal mammals become less active, presumably seeking out the shadows to conceal themselves from predators or prey. Our own closest relatives, the primates, buck this trend and seem to enjoy the increase in nocturnal light levels, as do some daytime predators such as wild dogs and cheetahs. Meanwhile, there are reports that lion attacks on humans are more likely to occur in the ten days after the Full Moon, perhaps taking advantage of the declining moonlight and also the fact that the waning Moon rises progressively later and later after sunset, making the early part of the night seem even darker to the daylight-adapted eyes of the lions' prey.

Out at sea, whole reefs of coral synchronize their spawning behaviour with the lunar phases, releasing their sperm and eggs around five days after the Full Moon. This seems to serve two purposes, maximizing the chances of fertilization by spawning all at once and taking advantage of the most favourable tides to disperse their offspring. Some species of sea turtles also time their egg laying according to the phases of the Moon, again to catch the most advantageous tides.

However, claims that the Moon can influence human biology have so far failed to stand up to rigorous scrutiny. For centuries, connections have been made between the human menstrual cycle and the phases of the

Moon: 'menstruation' comes from the Latin *mensis*, meaning '(lunar) month', and indeed the average length of the menstrual cycle is 28 days, which is quite close to the 29.5-day cycle of the lunar phases. However, among individual women, there is a large variation in the length of a typical cycle, ranging from 21 to 35 days, and careful studies have found no significant correlation between human menstrual cycles and the phases of the Moon, even in populations who live without artificial lighting and might therefore be expected to be more strongly affected by natural cycles of light and dark. It seems that the similarity in these biological and astronomical rhythms is probably a coincidence, but nevertheless in many traditions the Moon remains a potent symbol of fertility and a metaphor for female power.

Another lingering belief in the power of the Moon is the rather charming concept of 'moon gardening' in which various horticultural activities are timed to coincide with particular lunar phases in order to maximize vigorous plant growth. It is likely that this concept has its roots in the earliest days of human civilization, when the sky acted as a calendar for timing the key activities of the agricultural year. The practice continues to enjoy enthusiastic support in certain quarters to this day and some gardening and farmers' almanacs still include sections on lunar phases for moon gardening aficionados. Seeds sown just before the time of a Full Moon are claimed to enjoy a particular advantage, although carefully controlled studies have repeatedly failed to show any correlation between lunar phase at the time of planting and subsequent fruiting and flowering success.

Nevertheless, proponents of moon gardening have invoked various mechanisms to explain how the Moon might influence plant growth.

One claim is that the light of the Full Moon enables plants to photosynthesize at night, generating extra food to fuel their growth. But, aside from the fact that photosynthesis has little bearing on the germination of seeds under the soil, moonlight is still far too weak to be useful to plants – even the light of the Full Moon is around a million times fainter than sunlight. Another suggestion invokes the power of the lunar tides, arguing that since the Full Moon is also the time of the highest 'spring' tides in the sea, groundwater levels will also be raised, giving plants access to increased moisture. However, unlike the oceans, which span the entire globe and therefore feel the full difference in the lunar gravity from one side of the planet to the other, groundwater in the soil does not have a global distribution and is also not free to move around in the same way as seawater. Any conceivable influence of lunar tides on soil moisture in an individual field or garden would be utterly overpowered by other factors such as air pressure, humidity and the time since the last significant rainfall. If there is any effect at all, it is more likely to be a behavioural one: perhaps people with the patience to synchronize their activities with the lunar cycle are simply more thorough and careful than the average gardener.

The Moon may not determine our gardening success but it has still had a profound effect on the way that we understand our relationship to the universe around us. In 1609, Galileo's telescopic observations showed that the Moon was not, as had formerly been believed, a smooth globe, but was covered with rugged mountains – in other words a world in its own right, with landscapes not too dissimilar to those of the Earth. Just a few months later, he also discovered that the planet Jupiter had four moons of its own orbiting around it, just as the Moon orbits the Earth.

Both of these discoveries broke down the distinction between the Earth and the heavens: if objects that moved through space could have landscapes and moons just like the Earth, then couldn't the Earth also be an object moving through space? These ideas were at the heart of the Scientific Revolution, and helped to shape the technological world that we live in today.

In the twentieth century, the Moon also helped to redefine our relationship with the Earth itself. The race to land human beings on the Moon may have been driven by Cold War rivalries, absorbing a staggering 0.5 per cent of America's GDP for much of the 1960s, but its biggest legacy has been an international one. The famous 'Earthrise' images of our fragile blue planet hanging above the barren lunar horizon were a powerful reminder of humanity's shared heritage and the pictures are credited with helping to launch the global environmental movement. As Apollo 8 astronaut William Anders said, 'We came all this way to explore the Moon, and the most important thing is that we discovered the Earth.'

The Moon has profoundly affected the Earth in numerous ways over the last 4.5 billion years, but on an everyday basis its most obvious influence on our planet is surely that of the tides. The Royal Observatory was founded in order to provide lunar tables for use in maritime navigation, but sailors had already been living their lives according to the dictates of the Moon for centuries, because of its effects on the sea. Even in Greenwich, 50 kilometres from the open sea, the Thames is a strongly tidal river and twice each day enormous volumes of water surge up the estuary from the North Sea – enough to temporarily reverse the river's flow and to raise its level by up to 6 metres. Such powerful natural forces are not to be ignored: the height of the tide would determine which vessels could dock and where, while the rate of the tidal flow governed when ships would arrive in port or depart for destinations around the globe. Before the days of automated navigation, instruction in calculating the tides was an integral part of the education of officers graduating from the Royal Naval College on the banks of the Thames in the shadow of the Royal Observatory itself.

The ebb and flow of the tides seems like such an intrinsic property of the sea that it is easy to forget that the driving force behind these vast movements of water originates not here on Earth but far out in space. Indeed, they are powered by one of the most fundamental forces in the universe: gravity. As the force of attraction that acts between all particles of matter, we see the effects of gravity everywhere we look in the universe, from apples falling from trees to the motions of the largest galaxy clusters. Most of the time, it is only the familiar gravity of the Earth that we really notice – after all, this gravity keeps us, and everything else on our planet, stuck to the ground. But when it comes to the ocean tides, the gravitational forces at work are those of the Moon and, to a lesser extent, the Sun.

The tides are a complex phenomenon and, although people had linked them to the Moon for millennia, it is only in the last few centuries that we have come to understand how distant objects in space can actually affect the oceans here on Earth without resorting to mystical or astrological explanations. Perhaps this long confusion is not all that surprising. One of the main difficulties in understanding the origin of the tides is that, compared with the daily motions of the Moon and Sun across the sky, the rising and falling of the oceans can seem much more complicated. Once each day, the Sun and Moon rise in the east, reach their maximum heights in the sky, then decline again to set in the west. But, as anyone who spends time beside the sea will know, the tides are rarely this straightforward, and their pattern varies depending on exactly which coast you're standing on. While some areas of the world experience a 'diurnal tide', with only one cycle of high and low tide each day, elsewhere the waters peak twice a day and the tide is 'semi-diurnal'. Meanwhile, the times of high and low tides are not constant from one day to the next but change systematically, shifting later by just under an hour every day.

To confuse matters further, the height of the tides varies enormously from place to place around the world. Along some coasts the difference between high and low tide – the tidal range – can be many metres. The largest tidal ranges in the world have been recorded in the Bay of Fundy in Nova Scotia: here differences of over 16 metres have been measured between high and low tides. Even in Greenwich, far from the sea, the tidal range in the Thames is still impressive: at low tide the river is confined to a narrow channel between wide banks of gravel and sediment, but when the tide is high the muddy waters fill the riverbed from shore to shore, lapping just a few centimetres below the top of the embankment. Meanwhile other coasts are virtual strangers to any tidal variation at all: in some parts of the Mediterranean the daily difference is just a few centimetres.

But the tidal range is not fixed, even in a particular location; it also changes over time. Over a period of about two weeks, all around the world, the difference between the maximum and minimum height of the tides varies. Particularly high and low tides, known as 'spring' tides (in reference to their rapid rising and falling rather than the season – they occur throughout the year), alternate with periods of reduced contrast, known as 'neap' tides. Even these cycles of spring and neap tides can change in their extremity throughout the year, adding an extra layer of variability to the basic tidal pattern.

Despite all this complexity, careful observation reveals many telling similarities and synchronicities between the tidal motions in the sea and what goes on in the heavens – particularly the movement and phases of the Moon. Like the timing of the tides, the Moon's passage across the sky isn't synchronized with the Earth's 24-hour cycle of day and night: each day the Moon reaches its maximum height in the sky around 53 minutes later than the day before. From our modern perspective, we can understand this as a natural consequence of the fact that the Moon is moving in its orbit around the Earth, making a single complete circuit of our planet every 27.3 days. In each 24-hour period, therefore, the Moon moves about one-thirteenth of the way around its orbit: from our point of view down here on the ground, the Earth has to keep turning for an extra 53 minutes every day in order to bring the Moon back to the same place in the sky as it was yesterday.

In some regions, such as Maine on the east coast of the United States, a very obvious correspondence exists between the cycle of the tides and the Moon's passage across the sky. Here the tide peaks as the Moon reaches its highest point and then ebbs away once more as the Moon sinks towards the horizon. In most other places, the daily cycles of the tide and the Moon still shift in tandem but are offset from one another, so that there is a time lag or delay between the Moon reaching its zenith in the sky and the tide peaking in the ocean. On a longer timescale, it is also possible to see a correlation between the two-week cycle of spring and neap tides and the varying phases of the Moon, with the extreme spring tides coming just after Full Moon and New Moon, and the weaker neap tides coming around the times when the Moon is half full.

Despite speculation by astronomical heavyweights such as Kepler and Galileo, a proper scientific explanation of the tides did not come until the publication of Isaac Newton's great work, *Principia*, in 1687. Here, Newton laid out his theory of universal gravitation, partly inspired (according to a story told by Newton himself) by watching an apple fall from a tree in his garden. His theory proposed the existence of an invisible force – which he called 'gravity', from the Latin word for 'weight' – that operates between all particles of matter and acts to pull them closer together. Crucially the strength of this force falls off steeply with distance – in other words, the gravitational pull of an object gets weaker the further away from it you are. We call this particular kind of weakening with distance an 'inverse square law' because the strength of the gravitational force decreases with the square of the distance to the object: twice as far away from the object and its gravity is only a quarter as strong, four times further away and its gravity falls to one-sixteenth and so on.

This was a hugely significant achievement in the history of physics: Newton's theory was able to explain a whole range of different natural phenomena for the first time, from apples falling out of trees to comets orbiting around the Sun – and all using the same, simple, universal force. But it did not escape Newton's attention that, as well as accurately reproducing the motions of falling and orbiting objects, his gravitational equations also provided a very natural explanation for the tides. He realized that gravity provided the missing link, the invisible connection that allows distant astronomical objects like the Moon to influence the Earth's oceans, even across the empty gulfs of space.

According to Newton's theory, it is the gravitational attraction between the Earth and the Moon that binds them together. As we've seen, both objects are actually orbiting around their common centre of mass, but because the Earth is around a hundred times more massive than the Moon, the Earth's gravity dominates and the centre of mass of the Earth–Moon system lies within the Earth itself. But this isn't to say that the Moon's gravity has a negligible effect on the Earth. On the contrary, the Earth and everything on it feels the gravitational pull of the Moon – but, crucially, that pull is not the same everywhere on the planet. The Moon's gravity falls off with distance and the side of the Earth nearest to the Moon will feel a slightly stronger pull than the side furthest away. It is this gravitational difference that is at the root of the ocean tides.

It is not hard to imagine that a difference in the forces acting on two sides of a body like the Earth will inevitably have some kind of distorting effect. But the Earth is made of solid rock and, although the difference in the Moon's pull from one side to the other does cause a slight distortion in the shape of our planet's rocky crust, the effect is very small – about half a metre at the equator. However, our oceans are anything but solid: they're made of liquid water that is free to flow according to whatever gravitational influences are around. The Moon's gravitational tugging therefore gives rise to two great bulges of water on the Earth's surface: one on the side directly facing the Moon, where its pull is strongest, and one on the opposite side, where the Moon's pull is least.

Naturally, these watery bulges will always try to remain precisely in line with the direction of the Moon but the Earth is rotating on its axis faster than the Moon is orbiting around it. The bulges therefore don't keep pace with the surface of the Earth but lag behind as the planet spins beneath them. Of course, the seawater itself isn't fixed in place beneath the Moon; the oceans are still being carried around with the rotating surface of the Earth, but as one lot of water is rotated away from the bulge, another lot flows in to maintain it. From the point of view of a person at a fixed point on Earth's surface, the ocean therefore appears to surge up and down as the planet's rotation carries them through the tidal bulges of seawater.

This scenario gives us the basic pattern of two high tides each day. They are separated by around 12 hours and 26 minutes because the Moon itself isn't stationary. As the Moon moves along its orbit, every day the Earth needs to rotate for an extra

53 minutes or so to get back to the same position beneath the Moon and the corresponding bulge of the high tide.

But, as we've seen, not every shore experiences the simple situation of two high and two low tides per day, corresponding to the two tidal bulges of seawater, and many coasts experience only one high and one low tide in any 24 hours. Here, Galileo comes to the rescue. Despite the fact that he lived well before Newton's theory of gravitation – and that he didn't even believe that the tides were caused by the Moon at all – Galileo had correctly suggested that the timing and frequency of the tides would be modified by the local idiosyncrasies of shorelines and ocean basins. Depending on the shape of the coast and the depth and gradient of the sea floor, the simple, twice-daily tidal flow can be deflected, delayed, enhanced or even cancelled out entirely. This intricate complexity of local coastlines and sea bed is one of the main factors behind the variety of tidal patterns experienced around the world.

But the Moon isn't the whole story when it comes to the tides. Although its gravity is certainly the dominant force felt by our oceans, there's also another object out there with plenty of gravity of its own: the Sun. The Sun contains around 27 million times as much material as the Moon, with gravity to match, but despite this huge gravitational advantage its contribution to the Earth's tides is only around half that of the Moon. Once again the crucial factor is not so much the strength of the Sun's gravity as the amount by which it changes across the width of the Earth.

The Earth is around 12,000 kilometres in diameter, and this width is around 3 per cent of the overall distance between the Earth and the Moon – a fairly sizeable proportion. Because gravity decreases with the square of the distance, this means that the far side of the Earth feels a drop in the strength of the Moon's gravity of around 1 per cent compared with the near side.

By contrast, the Sun is around 150 million kilometres away. Even at this distance its overall gravitational pull on the Earth is around 170 times stronger than that of the Moon, but now the width of the Earth is tiny relative to the overall distance between the Earth and the Sun – only around a hundredth of a per cent. So the difference in the strength of the Sun's gravity from one side of the Earth to the other is just a ten-millionth of a per cent and its overall contribution on the tides works out at around half that of the Moon's. The result is that the Sun produces its own pair of tidal bulges in the oceans, smaller than those caused by the Moon but this time tracking the point on Earth directly beneath the Sun (and its opposite point on the far side of the planet).

As the Earth orbits around the Sun, the Moon orbits around the Earth, and the Earth spins on its axis, the Earth and Moon are both constantly changing their positions relative to the Sun and to each other. The constantly shifting directions of their gravitational interactions are responsible for much of the variation in the strength of the tides from day to day.

Roughly at New Moon, when the Moon and the Sun lie in the same direction in the sky, their gravitational pulls line up and their tidal bulges coincide. The same thing happens around two weeks later at Full Moon, when the Moon and the Sun lie on diametrically opposite sides of the sky. Once again, the Sun and Moon are aligned with the Earth, and their tidal influences reinforce each other. These are the times of the 'spring' tides when the high tides are highest and the low tides lowest. But halfway between New and Full Moon, when the Moon and the Sun are 90 degrees apart in the sky and the Moon appears half full, their gravitational pulls are at right angles and their tidal effects then work against each other, with the bulges from the Sun partially cancelling out those of the Moon. These are the neap tides, when the tidal range is reduced. It is this constantly varying competition between the lunar and solar influences that gives rise to the cycle of spring and neap tides – and this cycle naturally works in step with the changing phases of the Moon.

A further cause of tidal variation can be traced back to the fact that the Moon's orbit isn't perfectly circular – its shape is actually elliptical, or egg-shaped, with the difference between the Moon's closest and furthest approaches to the Earth amounting to more than 40,000 kilometres. When the Moon is at its closest to Earth, its pull is correspondingly stronger and, crucially, the *difference* in the strength of its pull across the width of the Earth also becomes slightly greater. This increased difference in pull from one side of the planet to the other produces a temporarily larger tidal range. When the Moon is further away, its pull is slightly decreased, and so is the difference, and the tidal range is therefore smaller than usual. This pattern is superimposed on the monthly cycle of spring and neap tides, with the result that during the course of the year some spring tides are more extreme than others.

A similar variation in the solar tides occurs due to the elliptical shape of the Earth's orbit around the Sun. For part of the year, the Earth is slightly closer to the Sun and the strength of the solar tides is therefore enhanced by a small amount. This variation doesn't change in step with the orbital variation of the Moon, leading to even more complexity in the overall range of the tides. Luckily, the variation in the strength of the solar tides is very small and can usually be ignored.

Even after Newton's great breakthrough, it took scientists and mathematicians over a century to explore all the nuances and complexities of the Earth's ocean tides completely, but at last they were in no doubt that gravity was the driving force. It is indeed the Moon and the Sun that, twice each day, force huge quantities of water up the Thames, then wrench it all back out to sea again. When we sit on the beach watching our sandcastles being swept away by the encroaching tide, we are witnessing the workings of gravity – a direct manifestation of this invisible force acting across the vast empty spaces of our Solar System.

But it is not only the Earth that is subject to tidal forces. Tides are also responsible for another very obvious astronomical phenomenon – the appearance of the Moon itself. Despite the ever-changing lunar pattern of light and dark as the Moon goes through its cycle of phases, one aspect of our natural satellite remains always constant: the Moon keeps one hemisphere permanently turned towards us. This means that from Earth we can only ever see one side of our celestial neighbour: the 'near side', with its familiar markings of dark 'seas' (actually smooth plains of ancient solidified lava) and lighter highlands, all pocked and scarred with innumerable impact craters. By contrast, the Moon's 'far side' is perpetually turned away from the Earth, forever hidden from view.

It is a common mistake to confuse the Moon's hidden 'far side' with the term 'dark side of the Moon'. In fact, the lunar 'far side' and 'dark side' are two separate things, which only sometimes happen to coincide with each other. Just like the Earth, at any one time, one half of the Moon is always in daylight while the other is experiencing night, and it is the half of the moon that happens to be experiencing night that is called the 'dark side'.

As with Earth, these regions of illumination and darkness are constantly moving around the entire globe, giving a day/night cycle, in this case to the Moon. From Earth, we see this as the changing lunar phases – a mix of illuminated and shadowed regions on the Moon, giving it the classic crescent and gibbous appearances. In all these cases, however, whether it's the daytime 'light side' basking in sunlight or the night-time 'dark side' hidden in shadow, the terrain we see is always, by its nature, the hemisphere that is facing the Earth – the 'near side' of the Moon.

In the same way, the far side of the Moon is also generally a mixture of lit-up and shadowed areas. At New Moon, when the side facing the Earth is entirely in shadow, the hidden far side is experiencing full daylight – so it's anything but dark. Only at

Full Moon, when the whole of the near side is illuminated and the opposite side is entirely in shadow, does the 'far side' of the Moon also become the 'dark side' and they are temporarily one and the same thing.

If the Moon's surface experiences a changing sequence of night and day, then surely, like the Earth, it must be spinning on its axis. Why then do we only ever see one side of it? The answer involves a piece of cosmic synchronization: the time it takes for the Moon to make one complete spin – a lunar day – is exactly the same as the time it takes it to make one complete orbit around the Earth – 27.3 Earth days. Effectively, as the Moon circles our planet it also twists on its axis by precisely the right amount to keep the same hemisphere always turned towards us. This synchronization seems far too neat to be a coincidence – and indeed it isn't. Once again, we're seeing direct evidence of tidal forces at work.

The idea of tidal forces acting on the dry and barren Moon might seem strange, but tides don't always require an ocean in order to make their presence felt. Just as the difference in the Moon's gravity from one side of the Earth to the other produces a distorting effect on our planet's liquid envelope, so the Earth's gravitational field exerts an even stronger distorting effect on the Moon. Despite its prominent 'seas' of dark, solidified lava, the Moon has no watery oceans to ebb and flow at the behest of its parent planet. But the Earth is about a hundred times more massive than the Moon, and its tidal pull on the smaller world is correspondingly more powerful – so powerful in fact that it can shape solid rock. The difference in the Earth's gravitational pull from one side of the Moon to the other has distorted its rocky outer layers, elongating our satellite along the direction extending towards and away from the Earth. Over millions of years, this tidal distortion away from a perfectly spherical shape has acted to slow down the Moon's rotation until it is now synchronized with its orbit – a process known as 'tidal locking'.

When the Moon first formed, it was probably spinning on its axis much more rapidly than it does today, and in the course of one orbital circuit around the Earth it would have presented both of its faces to our planet. But the tidal forces acting on the Moon created great bulges of rock in the Moon's outer layers and, just like the bulges of ocean tides on Earth, these rocky lunar distortions would try to track the position of the Earth as the Moon rotated.

However, if the Moon was spinning on its axis faster than it was orbiting the Earth, then it would constantly be trying to twist its tidal bulges away from their favoured positions facing directly towards and away from the Earth. But rock doesn't flow like water, so this unstable situation gave rise to a corrective, twisting force – a 'torque' – which tugs against this spinning motion, slowing the Moon's rotation in an attempt to keep the bulges aligned with the Earth. Similarly, if the Moon's rotation ever slows too much, so that the bulges now try to lag behind the position of the Earth, then the torque acts in reverse, spinning up the rotation of the Moon until the mismatch is minimized once again. The inevitable end result of all this is a Moon with a rotational period that is synchronized with its orbital period, so that its tidal bulges remain always lined up with the Earth. This means one hemisphere of the Moon permanently faces towards us while the other is permanently turned away.

It is astonishing to think that, for almost the whole of human history, no one had ever seen what the far side of the Moon actually looked like: a situation that persisted well into living memory. Fifty per cent of our satellite's surface was there, tantalizingly close, but always hidden behind the curve of the lunar horizon – providing a perfect stage set for science-fiction writers and conspiracy theorists and their most imaginative fantasies, but frustratingly inaccessible to scientific study and mapping. However, humanity's view of the Moon changed forever on 7 October 1959 when the Soviet space probe Luna-3 sent back the first grainy images of the mysterious lunar far side.

More than half a century and many space missions later, we now have detailed maps of the far side of the Moon, although still the only humans to have seen it with their own eyes are the crew members of NASA's Apollo programme, who flew around the far side as part of their mission to land on the lunar surface in the late 1960s and early 1970s.

What was immediately apparent to both Soviet scientists and Apollo astronauts was that the two sides of the Moon are actually surprisingly different from each other. Numerous impact craters scar both hemispheres, but there the similarity ends. The far side is dominated by ancient, pale grey highlands, with only a handful of the younger, darker 'seas' of solidified lava that give the near side its familiar mottled appearance. The reason for this two-faced nature is still not fully understood by lunar geologists but, like so many other aspects of the Moon, it seems very likely that it is also a consequence of those tidal forces. Perhaps, when the young Moon was still partially molten, the tidal stresses that caused it to synchronize its rotational and orbital motions also led to a thinning of the lunar crust on the side facing the Earth, allowing lava to escape more easily and producing the prominent 'seas' of the near side.

Whatever the exact explanation for the Moon's appearance turns out to be, even today tides continue to play a defining role in the evolution of the Earth–Moon system – a role that will profoundly shape their long-term future. The tidal influence of the Earth has already succeeded in synchronizing the Moon's rate of spin with its orbital speed. But that process is not yet over. Slowly but inexorably, the same tidal forces are working on the Earth, gradually reducing our planet's spin rate until it too is synchronized with the orbital period of the Moon. In the process, the Moon is also slowly spiralling away from us at a small, but measurable rate (*see* Written in the Stars, page 308).

The mechanics of this situation are very similar to those that caused the Moon to become tidally locked to the Earth millions of years ago. As the Earth spins rapidly on its axis, the two great tidal bulges of water try to remain massed directly beneath the slower-moving Moon while the Earth's rotation constantly tries to twist them away. Unlike the Moon's rocky distortions, the watery ocean tides of Earth can accommodate these conflicting pulls much more easily: as one lot of water is spun away, more simply flows in to replace it. But this compromise can't persist indefinitely because of another inescapable force: friction. As the oceans struggle to obey the tidal pull of the Moon, they act as giant brakes, dragging on the ocean floor and pushing against the continents – and this has the effect of gradually slowing the spin of the Earth.

The frictional dragging of the oceans means that the tides are actually making our days longer, stretching them out little by little as the Earth's rotation gradually winds down. At around 2 milliseconds every century, this slowing is imperceptible to human senses, but across the entire span of recorded history we find compelling evidence that the day has indeed been getting longer. Over the last few thousand years, the difference has been enough to noticeably affect the timing and location of certain types of astronomical event: historical records of eclipses over the centuries only make sense if the Earth's rotation is indeed slowing down.

Scientists can use the geological record to investigate how the length of a day has changed even more dramatically over the entire history of the Earth. Where layers of rock preserve sediments laid down by the rising and falling of the tides, these imprints of watery oscillations, known as 'rythmites', are a fascinating snapshot of the interaction between the Earth and the Moon in the distant past. But some of the most useful indications of the changing pattern of the tides are biological in origin: they come from fossils of living organisms, which can be sensitive to both daily variations in light and dark and the annual patterns of the seasons. Like tree rings, a record of these changes can sometimes be left in their growth patterns and be preserved if the organism is fossilized.

Such a record is preserved by some fossilized corals, which incorporate daily 'ridges' and annual 'bands' into their stony structures as they grow. By counting the number of ridges within each band, palaeontologists have calculated the ratio of daily to annual cycles. This tells us the number of times the Earth spins on its axis during one complete orbit of the Sun. Given that the Earth's orbit around the Sun has remained fairly constant since the formation of the Solar System, any change in this ratio will indicate that the length of a day must have been different from its current 24 hours. From the corals, palaeontologists deduce that, during the Devonian period around 400 million years ago, there were around 400 days in every year rather than the current 365.25, and a day was therefore two hours shorter than it is now. Going further back in time, microscopic organisms called cyanobacteria have left similar records in the form of stromatolites – mineral structures deposited layer by layer by the microbes over hundreds of years. From these, we can tell that 2 billion years ago a day would have lasted only around ten hours.

It is difficult to know how long a day would have been at every point in the Earth's history. The fossil record is naturally rather patchy and in any case the rate at which the Earth's rotation slows has probably not been constant. The strength of the braking process depends on the exact details of the tidal coupling between the Moon and the oceans; this is dictated by several factors including the distance between the Earth and Moon (which has gradually been increasing) and the pattern and strength of the ocean tides (which depend on the changing shape of the ocean basins as plate tectonics rearrange the continents). However, computer models indicate that when the Earth and

Moon first formed, around 4.5 billion years ago, our planet was probably whipping round on its axis once every six hours.

The gradual lengthening of the day due to the tides is one of the reasons why we occasionally need to add extra 'leap seconds' to the official time kept by atomic clocks. These timekeepers are accurate to one second in every 100 million years, using the oscillations of tiny electrons as a super-stable time standard. As the Earth's rotation continues to slow and our days gradually stretch themselves out, our atomic clocks will go on ticking out their perfect, unchanging seconds, minutes and hours. But we humans have a biology that is still tied to the rhythm of night and day and, without the addition of leap seconds, time according to the clocks would begin to lose its connection to this natural cycle of light and darkness.

As humans, we too are contributing in a tiny way to the lengthening of the day, when we use tidal barrages and turbines to generate renewable electricity from the ebb and flow of the tides. Such schemes impede the tidal flow to extract energy from the bulk movement of the water but this adds to the friction between the oceans and the Earth by a tiny amount and therefore contributes to the gradual slowing of our planet's rotation. By harnessing the tides we are actually extracting energy from the rotation of the Earth, although it is unlikely that our contribution will ever be noticeable.

It is not just the Earth whose motion is being altered by the tides. Gravity works both ways and, even as the Moon tries to hold the ocean water in place against the spin of the Earth, the tidal bulges are also exerting their own pull on the Moon. As the water spins around with the Earth, it tries to tug the Moon along with it, causing our satellite to gain energy and spiral out further and further from the Earth. Our planet is slowly losing its grip on the Moon as energy is removed from the Earth's spin and transferred to the orbital motion of the Moon, moving it further away.

If this process was allowed to continue to its natural conclusion, in the far future the Earth would finally settle into a tidal lock with the (by then) extremely distant Moon, each body keeping a single hemisphere permanently turned towards the other. With the Earth's day now the same length as a lunar month, the Moon would stand forever in the same place in the sky, visible only from the side of the Earth that faced it and undergoing a leisurely cycle of phases as it moved slowly along its vastly expanded orbit. Total solar eclipses would long since have become a thing of the past: the distant Moon would no longer appear large enough in the sky to completely cover the disc of the Sun.

In fact, a similar tidal scenario has already played itself out in the distant reaches of the Solar System. Here, the dwarf planet Pluto and binary companion Charon have been orbiting serenely in the frozen darkness for billions of years. Closer together and more similar in size and mass than the Earth and Moon, Pluto and Charon long ago became tidally locked to each other, their individual spin rates and orbital period now perfectly synchronized. They will hang, forever stationary in each other's skies for billions of years into the future.

However, despite the best efforts of the tides, this mutual locking will not be the ultimate fate of the Earth and Moon. As we shall see, long before our planet and its satellite can fully synchronize their motions, the Sun will warm and brighten as part of its natural evolution. In a couple of billion years, this warming will cause the Earth's oceans to evaporate, removing the main source of tidal coupling between the Earth and Moon and greatly reducing the exchange of energy from the spinning Earth to the orbiting Moon. In 5 billion years, the Sun will swell into a red giant star a hundred times its current size, very likely destroying the Earth and Moon in the process. Long before they ever have a chance to become fully tidally locked, the Earth and Moon will be gone forever.

The Moon is such a reassuring presence in the sky that it's hard to imagine that it is actually spiralling ever further from our planet's grip. Indeed, when the detailed mathematics of the process was first worked out in the nineteenth century it took a while for it to be fully accepted. But, strange though it may sound, we do have compelling evidence of the Moon's gradual retreat from a pleasingly simple technique known as lunar laser ranging. When NASA's Apollo astronauts left the Moon for the last time in the 1970s, they left behind on the lunar surface a series of large mirrors. These 'retroreflectors' are still used by scientists to bounce pulses of laser light off the Moon and measure the time each pulse takes to make a round trip back to the Earth. After more than four decades, their measurements are hard to argue with: each year the time taken for a pulse to travel to the Moon and back has got slightly longer. From this we can calculate that the Moon is currently receding from us at around

3.8 centimetres per year – roughly the same rate at which our fingernails grow.

Despite its slow retreat the Moon is unlikely to lose its status as humanity's favourite astronomical object any time soon. After decades of neglect by space agencies our natural satellite is back on the agenda as a prime destination for both manned and unmanned space missions and a successful moonshot has become a rite of passage for new spacefaring powers such as China and India, eager to prove their technological prowess. Scientists are also finding their interest in the Moon rekindled. Its rocks are likely to preserve a far more pristine record of the history of the Solar System than those of Earth – they may even hold clues to the formation of the Earth itself. Meanwhile, the Moon's far side, permanently facing away from our planet, could provide the perfect site for sensitive radio telescopes, shielding them from sources of electromagnetic interference on Earth.

The Moon may be a fickle source of light but it has been Earth's constant companion for almost the entire history of our planet. The close relationship between these two sister worlds has shaped them both, and it will continue to do so for billions of years to come.

On Stranger Tides

Another place in our own Solar System where tides play a prominent role is out among the moons of Jupiter. This giant world, 11 times the diameter of Earth and over 300 times as massive, commands at least 60 satellite moons – several of them comparable in size with our own Moon. Of these, two show some rather unexpected properties. Closest to Jupiter, Io is a seething ball of volcanic activity, its surface boiling with sulphurous eruptions and flooded with lakes of lava. Slightly further out, Europa is encased in ice, but its frozen surface is marked with a crazy-paving pattern of cracks, signs that below the icy crust lies an ocean of liquid water, perhaps 100 kilometres deep.

In both cases, the chaotic surface features of these moons reveal the presence of dynamic, fluid interiors: molten rock in the case of Io and liquid water in Europa. Together, they are among the strangest and most active bodies in the Solar System and yet by rights both should long ago have stopped showing signs of subsurface activity. Such small objects should quickly have lost any residual internal heat into the vast refrigerator of space causing their interiors to solidify. This is what happened to our own Moon billions of years ago, and is the reason why it is to a large extent geologically inert today. What could be keeping the insides of Io and Europa warm enough to remain liquid? The answer is, of course, tides.

Just like with the Earth–Moon system, Jupiter and its satellites are engaged in an endless gravitational dance – but because of Jupiter's huge mass the tidal stakes are even higher. Like our Moon, Io and Europa are tidally locked to their parent with one hemisphere always facing Jupiter and their shapes stretched along the direction extending towards and away from the planet. The moons are also a constant influence on each other and, as they move along their orbits, the changing gravitational pull of their neighbours ensures that their paths around Jupiter are never perfectly circular. Jupiter's tidal grip on the moons tightens and relaxes, repeatedly squeezing then releasing them, as they oscillate first towards then away from the planet. This continual tidal flexing causes the rocks of the moons' interiors to grind together, generating heat through friction.

These moons demonstrate once again the huge influence that tidal forces can exert. On Io, tides power some of the most violent volcanic activity in the Solar System, while on Europa they have created a subsurface ocean containing more water than all the seas of Earth combined. A similar tidal heating effect is believed to create pockets of liquid water beneath the ice of Saturn's tiny satellite Enceladus, and some scientists have suggested that the salty depths of Europa and Enceladus may be the most likely places in our Solar System to harbour extraterrestrial life. If so, there could be alien organisms that owe their very existence to the tides.

Tides can play both destructive and creative roles. One of the Solar System's most beautiful sights, the rings of Saturn, consist of billions of icy fragments in orbit around the planet, ranging in size from grains of dust to a few metres across. No one knows for sure how these astonishing structures formed but one suggestion is that they are the remains of a moon that strayed too close to its parent planet and was torn apart by tidal forces. As the moon drew closer and closer to Saturn, the difference in the gravitational pull on its near and far sides would eventually have become stronger than the moon's own self-gravity, causing it to disintegrate. But now, in a neat twist, it is probably also tidal forces that help to keep the rings stable: the repeated tugging of Saturn's other moons prevents the particles making up the rings from clumping back together, preserving these awe-inspiring structures for millions of years.

The destructive side of tides also became apparent in 1992 when Comet Shoemaker–Levy 9 swung close by Jupiter. Once again, the tidal strain of the planet's immense gravity was strong enough to tear the comet's frozen nucleus apart. Shattered into more than 20 fragments, some of them several kilometres across, the debris swooped in a great orbit around Jupiter until, two years later, it returned to rain down upon the planet, exploding on impact to produce a chain of immense scars in the upper atmosphere. But tidal forces aren't always conducive to spectacular cometary displays. In 2013, the much-hyped Comet ISON conspicuously failed to live up to early speculation that it might become the 'Comet of the Century', with a tail stretching across the sky and visible even in the daytime sky. Tumbling end over end through the intense gravitational field of the Sun, ISON's icy nucleus disintegrated under powerful tidal forces as it passed just 1.2 million kilometres above the solar surface – and the unlucky comet fizzled to nothing before the expectant eyes of the world's media.

Other cometary impacts may also owe their origins to tides. Far beyond the most distant planets, our Solar System is surrounded by a vast spherical shell of billions of small icy bodies, known as the Oort Cloud. Centred on the Sun, the shape of this cloud is stretched out by the tidal gradients of the Milky Way Galaxy, and this distortion may be responsible for sending the occasional object tumbling in towards the inner Solar System, where it becomes a comet. Such objects can pose a threat to our planet, and Earth's collision with a comet 65 million years ago may well have caused the extinction of the dinosaurs. This impact cleared the way for our mammalian ancestors to inherit the Earth, but a similar collision in the future could spell the end for human civilization – one more reason for us to respect the power of tides.

Tides are also in evidence in solar systems other than our own. Since the 1990s, hundreds of planets have been discovered circling stars other than the Sun, and we can now be confident that nearly every star in the sky has its own system of worlds in orbit around it. But although our own Solar System definitely isn't alone, that's not to say that it is entirely typical among planetary systems. Indeed, one of the biggest astronomical surprises of the last 20 years has been just how different some of these brave new worlds are from anything we see in orbit around our own Sun.

Planets in our Solar System fall into two broad categories. Orbiting close to the Sun, the small rocky planets Mercury, Venus, Earth and Mars have solid surfaces and relatively insignificant atmospheres. Much further out, the giant worlds Jupiter, Saturn, Uranus and Neptune are composed mostly of atmosphere, without a sharp distinction between gas and solid. But it turns out that planets around other stars don't always stick to these two neat divisions and among the most surprising are the worlds known as Hot Jupiters.

As their name suggests, Hot Jupiters are gas giant planets, like Jupiter and Saturn in our own Solar System. But whereas these familiar worlds orbit far from the Sun, Hot Jupiters are found incredibly close to their parent stars, often in orbits even tighter than that of Mercury. However, giant planets like this can only form far from their parent star, where conditions are cold enough for them to acquire large amounts of hydrogen and helium gas. In order for planets like this to be found so close to their parent stars, they must have moved there after they formed (*see also* Written in the Stars, page 308). Some of these planets are so close that they have become tidally locked to their star, with an axial rotation period identical to their year (the time it takes them to make one complete orbit). With one hemisphere bathed in permanent, scorching daylight and the other plunged into everlasting night, the weather on these strange worlds must be alien in the extreme, perhaps with howling winds transporting heat from the incandescent day-lit side to the frozen darkness of the opposite hemisphere.

Tides also operate on a galactic scale. Once referred to rather poetically as 'island universes', galaxies are normally quite remote from each other and, over billions of years, their relative isolation allows these immense collections of stars, gas and dust to evolve into beautifully symmetrical elliptical and spiral structures. The separation between galaxies is typically millions of light years, but as they drift through the vast expanses of intergalactic space they do occasionally come close enough for their gravity to influence each other and when this happens tidal effects are once again very much to the fore.

Galaxies are huge structures: the distance from one side to another can be tens of thousands of light years. So as they approach each other the stars and gas clouds on the galaxies' near sides will feel a much stronger pull from their neighbour than those on the distant far sides. This gravitational gradient begins to distort the galaxies' neat symmetric shapes, as stars and gas on the approaching sides begin to be pulled from their original orbits, adjusting themselves to the tidal forces in the same way that Earth's oceans flow to obey the pull of the Moon. As the interaction proceeds, the two galaxies become locked into a gravitational dance with each other, swinging round their common centre of mass while the tidal forces cause many of their constituent stars to be flung out into intergalactic space, forming long stellar streamers extending from the main body of the galaxy, known as 'tidal tails'. A famous and much-photographed example of this process is the Antennae Galaxies in the constellation of Corvus, an interacting pair with twin tidal tails stretching for hundreds of thousands of light years, giving the system the appearance of an insect with two long antennae. Eventually, these two galaxies will completely merge together and the stars in their tidal tails will fall back and be reabsorbed into the resulting giant galaxy. Other galaxy interactions are less drastic and, after a glancing encounter, the two galaxies will go their separate ways, although often still bearing the scars and distortions of their tidal tussle.

Perhaps unsurprisingly, the most powerful examples of tidal forces are found in the vicinity of some of the most extreme objects in the universe. As we've seen, when a massive star reaches the end of its life, it will destroy itself in a catastrophic explosion known as a supernova (*see* Made of Starstuff, page 6; Exploding Stars, page 319). Most of the star's outer layers are blasted violently into space but its innermost core is doomed to suffer a very different fate – quite the opposite in fact. Squeezed inwards by the explosion, then gripped by the irresistible force of its own gravity, the core will collapse into one of the most condensed objects in the universe: a neutron star or a black hole.

The exact fate of the core depends very much on the amount of material in the original star. For the majority of supernovae the remaining core will be massive enough for its gravity to crush it down into a neutron star: a sphere about 10 kilometres across, in which all the material of a star like the Sun is squeezed into a space no larger than a city. This violent compression is enough to crush the very atoms that make up the star, forcing their constituent electrons and protons together to form a jostling soup of neutrons – hence the name 'neutron star'. The material is so dense that a single teaspoon of it would contain the same mass as Mount Everest. This is an extreme state of matter, unlike anything we can create here on Earth. But in the very largest stars, the mass of the core is so large that even this ultra-condensed neutron soup can't hold itself up against its own gravity. For these stars, the collapse is unstoppable: the neutrons themselves are crushed into each other as the core collapses completely to a single point of infinite density. The star has become a black hole.

Black holes and neutron stars generate extreme gravity because of a combination of very large mass and very small size. This is because the strength of an object's gravitational attraction depends on two things: how much matter it contains and how close you are to it. A large star can be tens of millions of kilometres across, so although it contains a great deal of matter it also spreads this material over a very large volume of space. It is impossible to get close to all of the star's matter at the same time because even as you approach one part of the star most of the rest of it is still thousands of kilometres away from you. But by cramming a whole star's worth of material into a very small volume – just a few kilometres in diameter or less – neutron stars and black holes generate some of the most intense gravity anywhere in the universe. The tiny size of these super-dense objects makes it possible to get close to all of that mass in one go – and therefore to feel the combined force of all its gravity.

This extreme compactness also leads to extraordinarily powerful tidal effects on anything that gets too close. As always, the gravity of these objects falls off steeply with distance, and the closer you get to them, the more you will notice a difference in their gravitational pull from one side of an approaching body to another. If you were hanging, feet down, several kilometres above a neutron star or a black hole you would notice a distinctly stronger gravitational tug on your toes than on the top of your head. Closer still and this difference would start to be rather uncomfortable. Like a medieval prisoner stretched on a rack, you would feel yourself being elongated by the tidal difference between your head and your feet. Eventually the gravitational gradient across your body would become too much and you would be torn apart – giving a new meaning to the phrase 'rip tide'. For rather obvious reasons, the technical term for this process of extreme tidal disruption is 'spaghettification'.

Tides shape the wider universe around us as much as they continue to shape the world on which we live. From the unimaginable wrenching forces at the brink of a black hole's event horizon and the fiery volcanoes of Io to the rising and falling of the Earth's seas and the occasional leap second on our clocks, their influence is all around us. And tides may be even more intimately connected to us than this. With its rapid days and looming satellite, early Earth might sound almost like one of the alien worlds described in this chapter, but the fossil record also tells us that it was not long after this initial period that life first began on our planet. Could it be that tidal forces have played an important role in the formation and evolution of life itself – perhaps by nurturing early organisms in sheltered tidal pools, or providing a halfway house from which plants and animals could go on to colonize the land? We still have a great deal to learn about the origins and evolution of life on Earth but certainly on today's coasts the intertidal zones between high and low water are rich environments, where life seems to thrive on the regular influx of sunlight, seawater and nutrients. If key events in the history of life really did take place in these changeable regions between the land and the sea, then we ourselves might owe our very existence to the power of the tides.

Small Worlds

When the New Horizons probe was launched towards Pluto on January 19 2006 this ambitious, decade-long mission was heralded as the first spacecraft to target the ninth planet of the Solar System. But just eight months into its voyage New Horizons' billing had to be changed: Pluto was no longer considered to be the ninth planet, and the spacecraft was now heading towards an example of a new class of object, known as a 'dwarf planet'.

This act of astronomical rebranding was the result of a controversial vote by members of the International Astronomical Union in August 2006. This in turn had been precipitated by the discovery of 2003 UB313, an object around the same size as Pluto and slightly more massive, orbiting twice as far away from the Sun. It had become clear that Pluto was not alone on the edge of the Solar System and that there could in fact be dozens more such objects out there, waiting to be discovered. Should all of these tiny, distant worlds be classed as planets like Pluto or should they – along with Pluto itself – be given their own designation, one that better reflected their unique properties?

Since its discovery by American astronomer Clyde Tombaugh in 1930 Pluto had always been the odd one out of the planetary family. Set on a bizarrely tilted and elliptical orbit that sometimes brings it closer to the Sun than Neptune, Pluto resembles neither the rocky inner planets nor the outer gas giant worlds. It orbits in the Kuiper Belt, a region of the Solar System long suspected to harbour millions of small icy objects which, along with the even more distant Oort Cloud, formed the parent population from which comets were derived. By the late twentieth century, this picture was starting to be verified as, one by one, new objects such as Sedna, Orcus, Varuna and Quaoar were being found in the region of the Kuiper Belt. Smaller than Pluto, these were still quite substantial bodies, ranging in size from a few hundred to over a thousand kilometres in diameter. It seemed that Pluto might simply be one of the largest and nearest of a vast collection of icy objects, which shaded seamlessly into the Kuiper Belt's population of comet-sized snowballs. The discovery of 2003 UB313 was the final straw. First identified in 2005 (although the digital images which recorded it had actually been taken in 2003 – hence its name) by American astronomer Mike Brown and his team at the Palomar observatory, this object orbits even further out than the Kuiper Belt, in a related region of space known as the 'scattered disc'. It is comparable in size to Pluto and has 27 per cent more mass – so if Pluto qualified as a planet, then surely 2003 UB313 should too?

In fact, astronomers had already been through a similar debate back in the nineteenth century. In 1801 great excitement was caused by the discovery of Ceres, a new 'planet' orbiting the Sun between Mars and Jupiter. The relatively wide gap between these two classical planets had been a puzzle for some time and Ceres seemed to neatly fill the vacancy. But soon nagging doubts began to emerge. At around 950 kilometres across Ceres is much smaller than any of the other planets – less than a quarter the diameter of our own Moon in fact – and over the next few years other, similarly-sized objects – Pallas, Juno, Vesta, and many more – began to be discovered orbiting in the same region between Mars and Jupiter. At first these new discoveries were also classified as planets, but as the tally continued to increase it became clear that the objects inhabiting this part of space were somewhat different from the eight major planets of the Solar System and deserved a classification of their own. In the 1850s Ceres and its fellows were officially removed from the list of planets and over the next few years gradually became known as 'asteroids' (from the Greek for 'star-like') – a term first suggested by the astronomer William Herschel in 1802. At last the picture seemed clear: instead of a single large planet, the gap between Mars and Jupiter was filled with a belt of asteroids, ranging in size from hundreds of kilometres down to just a few metres. By 1868 a hundred asteroids were known and by 1981 the total came to around ten thousand. Today, with computerized surveys scanning the skies, we know of over a million, and the list is still growing.

In many ways Pluto's story mirrors that of Ceres and the asteroid belt. At the heart of the debate over its status in 2006 was a rather surprising question: what is a planet anyway? We might think that the answer is obvious: a planet is an object in orbit around a star, but which is smaller than a star and does not shine with its own light. But for some time this neat, simple picture had been getting increasingly confused.

In fact, our idea of what a planet is had grown up rather haphazardly, without ever seeming to need a very formal definition. Ancient people had noticed that there were several objects in the sky that constantly changed their positions against the constellations of the 'fixed' stars. These were the bright discs of the Sun and Moon, and the five star-like points of Mercury, Venus, Mars, Jupiter and Saturn, which the Greeks grouped together under the term *asteres planetai* or 'wandering stars'. In the sixteenth century Nicolaus Copernicus suggested that the Earth, rather than being fixed at the centre of the cosmos, was itself a wandering planet in orbit about the Sun, and when Galileo's telescope revealed four satellites moving around Jupiter in 1610, these miniature wanderers were also referred to at first as planets. Galileo's observations were instrumental in gaining acceptance for the Copernican model and gradually the terminology for describing this Sun-centred system settled into our modern concept of planets being objects that orbit the Sun, and moons being objects that orbit planets.

The number of known planets increased in 1781, when the musician and amateur astronomer William Herschel discovered Uranus from his back garden in Bath, and again in 1846 with the detection of Neptune, exactly where it had been predicted by the French mathematician Urbain Le Verrier, based on its gravitational effects on Uranus. These newly discovered planets were sufficiently similar to the six already known to arouse little controversy but by now the Solar System had become a much more complicated place. With his demonstration that comets moved around the Sun on elliptical paths Edmond Halley had already shown that planets were not the only objects in solar orbit, while the discovery of asteroids had added a whole new host of objects to the Sun's family. Moreover, the planets themselves seemed to be divided into two distinct types: the small, rocky worlds of the inner Solar System and the outer gas giant planets, Jupiter, Saturn, Uranus and Neptune. Some scientists even argue that distant Uranus and Neptune should be placed into a separate category of their own, known as 'ice giants'.

When Pluto was discovered in 1930 it seemed out of place almost from the start: a small world rolling through the frigid darkness at the edge of the Solar System. For half a century Pluto was believed to be somewhat larger than Mercury but in 1978 it was discovered to have a moon, Charon, whose orbit enabled Pluto's mass to be accurately determined for the first time. The result was surprising: Pluto is only 5 per cent the mass of Mercury and only around 2,400 kilometres across – smaller and less massive in fact than our own Moon. Moreover its satellite Charon was more than half as large Pluto itself – a situation unlike any other combination of planet and moon in the Solar System. (In fact Pluto is now known to also have at least four smaller moons – quite a haul for such a tiny object.)

To complicate matters still further, in the 1990s planets began to be discovered orbiting stars other than the Sun. The idea that the stars might have their own planetary systems around them dates back centuries, and the assumption had always been that they would closely resemble our own Solar System. But many of these 'exoplanets' turned out to be far stranger than anyone had imagined, with sizes, orbits and compositions unlike anything seen before, and it became clear that our own Sun's family did not represent the full range of objects that could be found in orbit around a star.

Into this increasingly confusing picture came 2003 UB313. The International Astronomical Union – the body officially responsible for assigning names and designations to celestial objects – assigned a task force of experts to report on the issue and on 24 August 2006, after a controversial vote, their recommendation was adopted as the IAU's *Definition of a Planet in the Solar System* (thus deferring the even more complicated issue of planets outside the Solar System for a future date). The definition includes three criteria that must be met in order for an object to qualify: a planet is a celestial body that is in orbit about the Sun, has sufficient mass for its own gravity to squeeze it into a roughly spherical shape and has cleared its own neighbourhood of rival objects.

Eight Solar System objects meet all three of these criteria: Mercury, Venus, Earth, Mars, Jupiter, Saturn, Uranus and Neptune. But tiny Pluto, although it orbits the Sun and has a spherical shape, falls at the final hurdle. Orbiting in a region full of other icy objects, some of which are of similar size and mass, it cannot be said to have cleared its neighbourhood – and so when the IAU voted to adopt the *Definition* they were also voting Pluto's planetary status away.

But, of course, the vote didn't remove Pluto's existence: if it wasn't a planet then, just as with the asteroids, it would require a new classification. The category created to fill the gap was that of 'dwarf planet', defined as a spherical object orbiting the Sun, but which has not cleared its neighbourhood of other objects.

Astronomers are not generally known for losing their temper but these decisions provoked huge controversy, with heated arguments being made both for and against the changes. Criticism was also levelled at the fact that only around 5 per cent of the astronomical community took part in the vote, although to be fair this still included a large fraction of the world's leading planetary experts.

However, for now at least, the decision stands. One of the first consequences of the new status was that 2003 UB313 – appropriately rechristened 'Eris' after the Greek goddess of discord – qualified as a dwarf planet alongside Pluto. Two other Kuiper Belt objects, Haumea and Makemake, joined the dwarf planet ranks and – perhaps settling a score dating back to 1801 – the largest asteroid, Ceres, also scraped in under the dwarf planet definition.

There's no doubt that the new definitions are still highly unpopular with many astronomers and, as we continue to discover more about the range of objects in orbit about the Sun, it remains to be seen whether the current categories will stand the test of time. Certainly there are many other objects – perhaps hundreds – in the outer Solar System that could qualify as dwarf planets so this is a category that may yet strain itself to breaking point. And the hundreds of planetary systems around other stars will doubtless have lessons of their own to teach us about the diversity of planets.

Journalists often use emotive terms like 'downgraded' and 'demoted' to describe Pluto's change of designation but it could equally be argued that, in becoming a prime example of a brand new class of objects, Pluto has achieved something of a coup. In

2006 New Horizons set out towards the smallest, coldest and most distant planet of the Solar System but in 2015 it arrived at one of the largest dwarf planets – a controversial object that had rekindled the curiosity of astronomers. On board the spacecraft are some of the ashes of Clyde Tombaugh, who forever changed our picture of the Solar System when he first spotted Pluto in 1930. Perhaps it is fitting that, like the Voyager and Pioneer probes before it, New Horizons is destined to speed on forever, leaving the Solar System and its controversies far behind.

Human beings like to classify things and place them neatly into boxes but perhaps the real message of both Pluto and Ceres is that the universe rarely conforms to our expectations, and our artificial categories are only an approximation to the bewildering and wonderful diversity of nature. There is one thing that we can be sure of: however we eventually decide to classify it, Pluto itself certainly doesn't care. It has already been there on the frozen rim of the Solar System for 4.5 billion years – long before we humans were around to know it existed – and it will still be making its long, cold circuits of the Sun a long time after we are gone.

Although Pluto and Ceres are currently classified together under the new fangled heading of 'dwarf planets' this doesn't negate their original affiliations: Pluto is still one of the largest known Kuiper Belt objects and Ceres is still the largest asteroid. As we've seen, both objects have fascinating tales to tell of the early stages of planet formation and of the origins of the Earth's water when the Solar System was still young, but it is their legions of smaller brethren – the comets and asteroids that are just a few kilometres across – which continue to influence our own planet 4.5 billion years later.

One of the biggest surprises in the images sent back by the New Horizons probe was the smoothness of Pluto's icy plains. In particular the relative lack of large craters indicated that the surface of the dwarf planet had to be young in geological terms – with an age of around 100 million years or so. This is because crater-gouging impacts with comets and asteroids continue to be a fact of life for every large body in the Solar System, even down to the present day.

Impacts have a long history: they were the mechanism by which all solid objects in the Solar System – planets, moons, dwarf planets, asteroids and Kuiper Belt objects – grew from the original dust particles of the pre-solar cloud, as particles collided and stuck together, building up into pebbles, then boulders, then mountains and finally into embryonic planets. The later stages of this process were spectacularly violent, with objects hundreds or even thousands of kilometres across raining down onto newly formed planets and moons. This age of titanic impacts is long gone and the Solar System settled into its current form around 4 billion years ago but, even today, not everything has been incorporated into the eight major planets and their moons. As the asteroid belt and the Kuiper Belt testify, there are still many trillions of smaller objects, from dust grains to boulders to objects several hundred kilometres across, in orbit around the Sun – and not all of them remain safely segregated in neat, well-constrained zones between Mars and Jupiter or out beyond Neptune. When such wandering objects stray too close to a planet or moon, violent impacts can still occur. The Solar System may have calmed down significantly since the violence of its youth, but the planet-building process has never completely stopped – it still continues at a low level in the form of these occasional encounters.

Everywhere we look in the Solar System the solid surfaces of planets, moons and other objects are pocked by craters, the scars of impacts with comets and asteroids and smaller pieces of rocky and icy debris, which have taken place over the last 4.5 billion years. Only where geological processes regularly renew a planetary surface – as on Pluto, or indeed here on Earth – is the record of these impacts obscured or removed. But it's important for us to remember that a lack of visible craters does not imply that a world is immune to the threat of random collisions. The pockmarked surface of our own Moon – a complex mosaic of craters upon craters, many of them ancient but others alarmingly recent – is a sombre reminder that the Earth remains very much in the firing line.

Indeed it turns out that our planet receives around 40,000 tonnes of extraterrestrial material every year, mostly in the form of interplanetary dust grains but occasionally in significantly larger chunks. On a typical day at least one object around 40 centimetres across will strike our planet, while each year several metre-sized objects will also arrive – and on timescales of decades and centuries we can expect much larger visitors. Happily, unlike the Moon or even Pluto, the Earth is not entirely without natural defences. The atmosphere may not seem like much of a barrier to us but objects in orbit around the Sun are typically moving at several kilometres per second and, at such high speeds, hitting a wall of air can be a traumatic experience. As the impacting object forces itself through the atmosphere the air in front of it is violently compressed and heated; the object streaks through the sky enveloped in an incandescent sheath of gas, appearing from the ground as a meteor or 'shooting star'.

Meteors generally become visible at heights of around 100 kilometres above the ground (appropriately, since their name comes from the Greek word *meteoros* or 'high in the air') and the temperatures generated in the air around them can be greater than 1,500° Celsius – hot enough to begin to vaporize the ice, rock or metal of the object itself. Objects smaller than a few centimetres across are likely to burn up entirely before they ever reach the ground and in this way the Earth's atmosphere protects us from the majority of these tiny missiles. Look up on any clear night for long enough and you will have a good chance of seeing a meteor streak across the sky – a fiery end for a particle that had been floating between the planets for billions of years. At certain times of year, when the Earth passes through the trails of dust left along the paths of passing comets, we are even treated to meteor showers – cosmic firework displays that can occasionally produce hundreds of meteors every hour.

Surprisingly, the smallest dust grains don't burn up at all since they are too small to significantly compress the air around them. Instead they are gently decelerated and sift down to the surface – a constant rain of interplanetary particles. Larger objects – typically those with sizes of tens of centimetres or greater – can also survive the journey, albeit with the loss of their outer layers, and when they reach the surface they become known as meteorites. As samples of material from elsewhere in the Solar System they are highly prized by planetary scientists and a whole discipline known as meteoritics is devoted to their study – indeed, they are one of our few means of directly investigating the composition of objects in space without actually going there (*see* Alien Oceans, page 75).

Scientists will go to great lengths to acquire new meteorite samples, mounting gruelling expeditions to the Sahara Desert or the ice sheets of Antarctica where even objects that arrived centuries ago still lie undisturbed where they fell, their burnt and blackened exteriors conspicuous against the sand or ice. Even better than collecting ancient falls is to actually see a meteorite hit the ground. By collecting it straight away you can minimise the amount of contamination that it receives from its terrestrial environment, thus keeping it as pristine as possible, while the angle and direction of its inbound trajectory can even provide information on where in the Solar System it originates.

However, when objects of metre-size or larger enter the atmosphere, being in their vicinity is probably not such a good idea. The energy carried by a moving object is proportional to its mass – in other words how much material it contains – and also to the square of its speed. This means that even an object of relatively modest mass can carry a large amount of energy if it's moving quickly enough – and the same object travelling twice as fast will carry four times as much energy. On entering the atmosphere some of this energy is dissipated as heat, forming the glowing trail of a meteor and driving a shockwave through the air, but if the object survives to strike the ground the remaining energy is released explosively on impact in an intense flash of heat and light and a blast wave that spreads rapidly outwards, pulverising rock to form a crater and displacing vast amounts of air in all directions.

On 15 February 2013 the residents of the Russian city of Chelyabinsk had a terrifying demonstration of the power that such an impact can unleash. At around 09:20 local time a small asteroid, about 20 metres across and weighing around 12,000 tonnes, entered the Earth's atmosphere at a shallow angle and raced westwards across Russia at a speed of around 19 kilometres per second – about 60 times the speed of sound. Blazing more brightly than the Sun the object was visible for miles around and its passage through the sky was caught on CCTV as well as the dashboard cameras that are commonly mounted in Russian cars for insurance purposes, making this one of the most documented impact events in history. At a height of about 26 kilometres the asteroid exploded, releasing more than 20 times the energy of the atomic bomb detonated at Hiroshima. Small fragments of rock rained down over a wide area but the greatest damage was done by the shockwave generated by the explosion. The dazzling spectacle had brought hundreds of residents rushing to their windows and when the blast arrived minutes later they were showered with broken glass. Nearly 1,500 people were injured and some also reported being temporarily blinded by the flash, while more than 7,000 buildings were damaged at an estimated cost of around $30 million. Most disturbingly of all, although radar systems are perfectly capable of detecting objects of this size at quite some distance from the Earth, no one had seen this asteroid coming – because no one had been looking for it.

But the Chelyabinsk event was not the most devastating asteroid impact of the twentieth century, and it is only by good fortune that the event that holds this record occurred over an unpopulated area. On the morning of 30 June 1908 an object estimated to be around 60 metres in diameter – three times larger than the Chelyabinsk meteorite – exploded above the pine forests of the Tunguska River in Siberia with a blast equivalent to 1,000 Hiroshimas. Like Chelyabinsk, the Tunguska object exploded in mid air, leaving no crater, but even so its effects on the ground were devastating. Heat from the explosion set fire to the forest and the shockwave flattened trees across a region of over 2,000 square kilometres, shattering windows several hundred kilometres away and even registering on seismic detectors in Western Europe.

Scientists estimate that impacts on the scale of the Tunguska event probably occur every few hundred years, but 60 metres is quite a modest size by the standards of asteroids and comets, many of which have sizes of several kilometres or more. Could such an object ever strike the Earth, and what would happen if it did?

The answers to both these questions have been found not by looking up into space, but by looking down at the rocks beneath our feet. For 135 million years the Earth's land-based ecosystems were dominated by the dinosaurs, an incredibly successful group of animals that had diversified into a huge variety of species, including some of the largest creatures that have ever lived. Their fossils are abundant in rocks from the beginning of the Jurassic period to the end of the Cretaceous 66 million years ago and then, quite suddenly, all trace of them disappears from the geological record. Something drastic must have happened to end their long reign, and in the 1980s scientists discovered several startling pieces of evidence that pointed to a catastrophe from the sky.

All around the world, at the boundary between the rocks of the Cretaceous period and the younger rocks of the Palaeogene, is a thin band of sedimentary rock containing an unusual abundance of the metal iridium. Iridium is generally rare in the rocks of the Earth's crust but it is often common in meteorites, and this suggested that a giant impact had occurred somewhere on Earth, spreading extraterrestrial debris around the globe. Also found embedded in the boundary layer are tiny crystals of fractured quartz – and since quartz is an extremely tough mineral, these too are evidence of an event involving extremes of heat and pressure. The smoking gun was identified in 1990, in the form of huge impact crater buried under layers of sedimentary rock at Chicxulub on the coast of Mexico's Yucatan peninsula. The crater is 180 kilometres across and it was formed 66 million years ago.

It's hard to say for sure whether the impacting object was an asteroid or a comet but from the size of the crater at Chicxulub we can tell that it must have been 10 to 20 kilometres across. Travelling at several kilometres per second the object vaporized on impact in an incandescent pulse of heat and light, blasting debris into space and sending a hurricane-force shockwave racing around the planet. The intense flash of the explosion ignited forest fires for thousands of kilometres in all directions, while the larger pieces of debris rained down across the globe triggering further Tunguska-scale explosions and igniting yet more fires.

Meanwhile, the shockwave rolled across the neighbouring land-masses of North and South America, flattening everything in its path. Because the impact happened by the sea, vast quantities of water were displaced, sending tsunamis hundreds of metres high crashing into coastlines around the Caribbean. Within hours, everything within thousands of kilometres of the impact site had been reduced to a wasteland of ash and mud. But the worst was yet to come.

Dust from the impact encircled the globe, spreading throughout the upper atmosphere and blotting out the Sun. Down on the ground, darkness and cold reigned for months on end and, deprived of sunlight to power photosynthesis, global food chains collapsed as plants began to die. Eventually, after perhaps a year of near-constant darkness, the skies began to clear but by then it was too late: 75 per cent of all the Earth's species, including the dinosaurs, had already died out in one of the greatest mass extinctions in the history of life on Earth. For the lucky 25 per cent of species that survived the explosion and the subsequent 'impact winter', the empty Earth presented a world of new opportunities, allowing them to diversify and flourish in their turn. Among them were the birds – the only group of dinosaurs to ride out the disaster – and our own mammalian ancestors.

From the beginning of the Solar System impacts have always been a double-edged sword. Without them there would be no planets at all and, as we saw in the chapter Water Everywhere, it may well have been the impacts of comets and asteroids that delivered this life-giving substance to Earth. Arguably the extinction event at the end of the Cretaceous could also be seen in a positive light – without it the dinosaurs might still dominate the Earth, and we would not be here at all. But another impact on this scale would not be such good news for us if it occurred today. The impact itself and the subsequent shockwave, fires and tsunamis would doubtless kill millions of people within hours, while hundreds of millions more would perish during the months of darkness as global agriculture collapsed and food stocks dwindled. Our species would probably not face total extinction but civilization would be crippled and those who survived would face a devastated world whose ecosystems would take tens of thousands of years to recover.

The potential impact of a large asteroid or comet is therefore one of the most serious natural disasters facing our planet today. We know that such impacts have happened in the past and, although they are rare, occurring on timescales of tens of millions of years, they will inevitably happen again in the future. However, this is one disaster that we have the power to avert. Unlike the dinosaurs, over the last few decades we humans have developed the technology not only to detect dangerous objects while they are still far out in space but also to attempt to divert or destroy them before they ever reach us.

Spurred on by a combination of warnings from scientists, sensational sci-fi disaster stories and real-life events such as the Chelyabinsk impact, governments are at last beginning to take the impact threat seriously and there are now several dedicated schemes to map the heavens and identify any potentially hazardous asteroids and comets. Reassuringly, the hunt has so far turned up no objects larger than a kilometre in size that could pose a threat to the Earth within the next few hundred years – which is about as far ahead as we can reliably predict their orbits. The focus is now turning to the smaller but far more numerous objects – those of a few tens to a few hundreds of metres in diameter – which, though unlikely to cause a global extinction, could still wipe out an entire city.

If and when we do discover an object on a collision course with our planet what could we do about it? The Hollywood solution would perhaps be to send a fleet of nuclear missiles to blow it to smithereens, but this might not always be the best course of action. In 2005 NASA's Deep Impact mission sent a washing-machine-sized probe on a collision course with Comet Tempel 1, excavating a 150-metre wide crater and releasing a huge cloud of vapour and dust from beneath the comet's icy surface. But this was not simply an act of cosmic vandalism. The impact proved what astronomers had already suspected: that comets are fragile structures of loosely bound material, only weakly held together by their own gravity. Apply too much brute force to such an object and you might only succeed in breaking it into thousands of fragments – many of which could still be travelling towards the Earth. It's not entirely clear whether thousands of Tunguska-style impacts distributed over a wide area would be less damaging than the impact of a single kilometre-scale object.

Instead astronomers and engineers have proposed a number of ingenious but less brutal strategies for deflecting hazardous objects. After all, from a great distance the Earth presents a relatively small target in the vast emptiness of space; if we catch an incoming object early enough, with decades or centuries to spare, then we will only need to divert its path by a very small amount in order for it to miss our planet entirely. Among the schemes suggested are attaching a rocket motor to steadily drive the object into a different orbit, or using mirrors to focus sunlight onto the surface, causing jets of vaporized ice or rock to erupt into space and pushing the object in the opposite direction. Gentlest of all is the proposal to spray the offending comet or asteroid with shiny paint and then sit back and let the infinitesimal pressure of sunlight reflecting from the surface gradually move it out of the way.

But the key to defending the Earth from the threat of extraterrestrial impacts is knowledge – not just of what's out there in space but also of what it's made of and how this material behaves under different conditions. Understanding the universe around us is never just an exercise in idle curiosity. In the end, the most important difference between the dinosaurs and us might turn out to be that we have telescopes and they did not.

And Yet It Moves

Unless you're reading this on a train, plane or other form of transport, you're probably under the impression that you're sitting perfectly still. But in fact we're actually in constant motion even when we think we're not moving at all – and at speeds that are quite dizzying to contemplate. We are living on the surface of a spinning planet, orbiting a star, which is moving through a galaxy, which in turn is travelling through space under the gravitational influence of the other galaxies, gas and dark matter that make up the material universe.

Discovering that the solid ground beneath our feet is actually rushing through space has profoundly changed the way we think about our relationship to the universe. We certainly can't claim that our world occupies a privileged central position, with everything else in the cosmos moving around it. Moreover, we now understand that the various motions of the Earth play a vital role in shaping the world that we inhabit. If our planet wasn't moving, its atmosphere, oceans, geology and climate would all behave very differently – in fact, it's likely that a stationary Earth would not be hospitable to life at all.

Yet, despite all this hectic motion, down here on the surface it doesn't actually feel as though we're going anywhere – while, in contrast, when we look up into the sky we see many objects that definitely do seem to be moving. Every day the Sun, Moon, stars and planets all appear to make a circuit of the Earth, while on longer timescales of weeks, months and years, the Sun, Moon and planets also move relative to each other and to the constellations of stars. Of course, we now know that many of these apparent celestial motions are actually due to the movement of the Earth, but for thousands of years this idea seemed absurd. Not only did it go against all common sense and everyday experience, but it also challenged deeply held philosophical, religious and even political beliefs about humanity's place in the universe. For astronomy to embrace the once-absurd idea took thousands of years, in a journey that reveals humankind's changing sense of itself and its own importance.

Lacking telescopes or other technological aids to vision, the astronomers of antiquity had to rely on what they could see in the heavens with their own eyes. However, they were highly sophisticated observers, mathematicians and philosophers and, by combining their observations with deductions based on reason and logic, they were able to come up with a model of the cosmos that seemed to match the behaviour of the sky.

At the centre of everything – solid and unmoving – was the Earth. Around it, the Moon, Sun and planets revolved and, since unsupported objects would surely fall to the ground, they were attached to a series of transparent crystal spheres that were nested one around the other. Surrounding them all was the sphere of the fixed stars, which contained the unchanging constellations. Each sphere revolved at a slightly different rate, giving rise to the gradual movement of the Sun, Moon and planets against the backdrop of the stars. Not only did this Earth-centred 'geocentric' model explain why everything appeared to move around us, it was also intellectually satisfying since it accorded with the prevailing philosophy of the time: things on Earth were transient, imperfect and corruptible but the heavens were made of altogether different stuff – eternal, unchanging and pure. So it was natural that all motion there would be based on the most perfect shape of all: the circle.

The geocentric model had the backing of some (though, as we shall see, not all) of the greatest philosophers of the ancient world and, despite its pagan origins, in due course it was also enthusiastically embraced by the Catholic Church, for whom it came to symbolize the hierarchy of divine and secular powers on Earth. God in His heavenly perfection ruled over all, while below Him were the Church, then the Earthly authority of emperors, kings and the nobility, and finally – at the bottom of the heap and decidedly imperfect – were the common folk.

In 1633, the Italian astronomer and mathematician Galileo Galilei was found guilty of heresy by the Roman Inquisition and sentenced to house arrest. His crime was advocating the idea that the Sun, not the Earth, was at the centre of the universe, and that the Earth was in constant motion around it. In 1616, the Church had declared this Sun-centred or 'heliocentric' model to be 'foolish and absurd' and, worse, to be in direct conflict with Holy Scripture. After all, according to the Psalms, 'the Lord set the earth on its foundations; it can never be moved', while Ecclesiastes states, 'And the sun rises and sets and returns to its place'. And in the Old Testament story of Joshua, the biblical hero commands the Sun to 'stand still' over the city of Gibeon – surely implying that it was the Sun, rather than the Earth, that was moving? Galileo had countered this criticism by arguing that the biblical accounts were meant to be poetic or allegorical but this cut no ice with the Inquisition and he had been ordered not to teach or defend the heliocentric model in public. Galileo's crime in 1633 was therefore one of disobedience as well as heresy, but implicit in it all was a direct challenge, not just to the intellectual authority of the ancient philosophers, but also to the established social, religious and political order of Europe.

Under threat of torture, Galileo publicly recanted the notion that the Earth is in motion around the Sun, but as he was led away from the courtroom popular legend has him defiantly muttering the phrase 'eppur si muove' – 'and yet it moves'. Most historians agree that there is little evidence that he actually said this, although Galileo still clearly continued to believe that the Earth did indeed move right up until his death nine years later. Nevertheless, the tale has come to symbolize defiance in the face of intellectual oppression and, like all such stories, it has been adapted through the ages to reflect the intellectual concerns of the day – perhaps most famously by Bertolt Brecht whose play *Life of Galileo*, written in the wake of the Manhattan Project and the atomic bombs dropped on Hiroshima and Nagasaki, used the story to highlight the responsibility of scientists to stand up to political pressure.

In fact, the idea of a heliocentric universe was not a new one, even in Galileo's lifetime. In 1543, the Polish astronomer Nicolaus Copernicus had published *De Revolutionibus orbium coelestium* (On the Revolutions of the Heavenly Spheres), in which he set out his ideas about the heliocentric model in detail, removing the Earth from its privileged position at the centre of the cosmos and replacing it with the Sun. The Earth was relegated to the status of a planet, orbiting the Sun along with Mercury, Venus, Mars, Jupiter and Saturn and third in order from the centre. Surrounding them all were the fixed stars, which – like the Sun – did not move. As well as orbiting around the Sun, the Earth was also spinning on its axis every 24 hours, explaining the daily revolution of the sky. In Copernicus' system, the only celestial body that still moved around the Earth itself was the Moon – the last legacy of the geocentric system. But Copernicus himself acknowledged that the idea of a Sun-centred cosmos went back still further, to the third century BC and the work of the ancient Greek astronomer Aristarchus of Samos.

Like many other texts from classical antiquity, Aristarchus' own works have not survived and we know of his ideas only through the writings of later authors such as Archimedes. From the remaining evidence, it seems as though the Greeks took his theory quite seriously – at least in principle – but in reality there were aspects of Aristarchus' heliocentric model that made it hard for the practically minded Greeks to reconcile it with everyday experience. For instance, Aristarchus' contemporary the mathematician and poet Eratosthenes of Cyrene had made a remarkably accurate estimate of the circumference of the Earth, to within a smal percentage of its true value of around 40,000 kilometres. In order for the Earth to spin once on its axis every 24 hours it must be rotating at an enormous speed – and yet we feel no effects from this furious motion. Why, for example, did this constant rushing around not create hurricane-force winds that would flatten everything in their path? In our modern world of high-speed transport, we're used to the idea that the air inside a train or a car is moving with the vehicle and so it isn't hard for us to understand that the Earth's atmosphere spins along with the planet. In fact, as we'll see later on (*see* Spinning Around, page 265), the Earth's spin does have a profound influence on winds and air currents, but not quite in the way that the Ancient Greeks imagined.

Another objection was that if the Earth was indeed moving around the Sun we should see evidence of our changing position in the form of parallax. The effects of parallax are familiar to anyone who has watched the changing view from the window of a moving train: nearby objects such as trees and buildings appear to move in front of the more distant backdrop of hills and horizon as our own viewing angle alters. Similarly, when we move closer to an extended object it looks larger and covers a greater fraction of our field of view. By the same logic, as the Earth moves around the Sun we would expect the constellations of stars in different parts of the sky to loom and recede as we periodically approach them and then move away. But, try as they might, ancient astronomers could not measure any difference in the apparent positions or appearance of the stars throughout the year – and this in turn seemed to argue against the movement of the Earth. Of course, if the stars are extremely far away, then the apparent change in position will be correspondingly small, but the Greeks reasoned that in order for it not to be detectable at all the stars would have to be stupendously remote – lying far beyond even the most distant of the planets. Was it really plausible to think that the universe could contain so much empty space?

Again, we now accept that the stars really are astoundingly far away: the distance to Proxima Centauri, our nearest stellar neighbour, is more than 200,000 times greater than the distance between the Sun and the Earth, and most stars are considerably further away than that. At such huge distances, the annual parallax of the stars due to the Earth's motion around the Sun is tiny: less than an arcsecond, or a sixtieth of a sixtieth of a degree. This is far too small to detect with the human eye and even once the telescope had been invented it wasn't until the nineteenth century that the technology had improved to the point at which stellar parallax could be measured.

In the face of these then quite reasonable objections, Aristarchus' heliocentric model seemed less credible than the geocentric alternative – in which, because the Earth didn't move at all, there was no need to worry about gale-force winds or looming constellations. Geocentrism was championed by some of the most formidable names in classical philosophy and astronomy, including Aristotle and Ptolemy, but even so the Greeks were painfully aware that the geocentric model had problems of its own. In particular, the observed motion of the planets refused to conform exactly to the perfect circular motion required by the crystal spheres.

Everything in the sky appears to circle the Earth once every day, but over timescales of days, weeks and years the Sun, Moon and planets also appear to move against the backdrop of the constellations. In fact the name 'planet' comes from the Ancient Greek term *asteres planetai* or 'wandering stars', precisely because of this motion with respect to the 'fixed stars' of the constellations, which never seemed to change their positions relative to each other. In the geocentric model this was explained by having each of the crystal spheres revolve at a slightly different rate with respect to the outermost sphere of the stars. They would all turn around the Earth roughly once per day, but the inner spheres turned slightly more slowly than the outermost one, and so gradually the Sun, Moon and planets would slip eastwards with respect to the stars, each at its own particular pace. For example, the sphere carrying the Moon would make a complete circuit in just one month, while the sphere carrying Saturn – for the Greeks, the outermost of the known planets – would take 29.5 years to return to the same configuration with the outer sphere. Because the spheres were perfect in their geometry the planets would trace perfectly circular paths around a ring of constellations that girdled the sky, known as the Zodiac (a name that comes from the Greek for 'Circle of Animals', so-called because many of the 12 constellations in the ring – Taurus the Bull, Leo the Lion, Pisces the Fish and so on – were named after living creatures).

Unfortunately, this neat model doesn't exactly mimic how the planets actually behave in practice. On the whole, they do move dutifully eastwards around the Zodiac but, every so often, Mars, Jupiter and Saturn deviate from their paths. Their eastwards motion slows, then reverses and, over the course of several weeks, they perform a little backwards loop in the sky before resuming their eastwards course again. Careful observation also revealed that the planets did not move at a constant speed against the background stars but appeared to speed up and slow down. The circular movement of the crystal spheres could not account for these changes in speed, or for the occasional loops of 'retrograde motion', and generations of astronomers struggled to find a plausible explanation.

In AD 150, using centuries worth of previous astronomical observations, Claudius Ptolemy of Alexandria came up with a mathematical modification to the geocentric model that seemed to wrench it back into alignment with reality. His ingenious solution used circles within circles to explain the occasional retrograde deviations of the planets. As well as being carried around by the motion of the crystal spheres, Ptolemy proposed that they were also looping around on smaller circles known as epicycles. Adding an extra level of complexity to Ptolemy's idea was the fact that, in order to fully reproduce all of the observed deviations from a purely circular path, the centre of the epicyclic motion had to be slightly offset from the Earth. Ptolemy's model was able to reproduce the actual motions of the heavens with tolerable accuracy and it remained in use for another 1,500 years. However, by remaining faithful to the philosophical ideals of a central, unmoving Earth and a heaven of perfect circular motions, Ptolemy had produced a model that was both complicated and rather arbitrary. Worse still, the complex motion of circles within circles made calculating the future positions of celestial objects a time-consuming chore.

When Copernicus resurrected Aristarchus' Sun-centred model in 1543, one of the main points in its favour was that it contained a natural explanation for retrograde planetary motion. If the Earth was a planet, moving around the Sun along with all the other planets, then our own motion could be responsible for their apparent deviations. As the outer planets Mars, Jupiter and Saturn have further to travel in order to complete one orbit, the Earth must regularly overtake them as it moves around its shorter circuit, rather like a runner on the inside lane of a circular race-track overtaking her competitors in the outer lanes. It was simple to show that, as Earth catches up with and then overtakes its neighbours, from our perspective they will appear to perform a brief backwards loop in the sky before resuming their former course – even though in reality they never deviate from their own orbits around the Sun.

Copernicus' desire to set the stationary Earth in motion seems to have come as much from Renaissance ideas of harmony and simplicity in nature as an urge to overturn centuries of received wisdom: the explanatory power of the heliocentric model was more aesthetically appealing than the complicated wheels within wheels of Ptolemy's geocentrism. However, Copernicus was not unaware of the idea's revolutionary potential and, perhaps because he feared criticism, he elayed publishing his thoughts on helio-centrism until the very end of his life: he died just as his book *De Revolutionibus* went to press. But, although they provoked much debate and argument, Copernicus' ideas avoided official condem-nation by the Catholic Church for over 70 years. Instead, the authorities focused on a more practical point: by assuming that the Earth was in motion about the Sun it became much simpler to calculate the future positions of celestial bodies. This was a great boon to the Church, which relied on calculations of lunar phases and equinoxes to determine the dates of important religious festi-vals such as Easter. Indeed, in 1582, Pope Gregory XIII used Copernicus' model to reform the calendar system, giving us the

Gregorian calendar that we still use today. It seems that it was fine to pretend that the Earth revolved around the Sun for the sake of mathematical convenience, as long as nobody actually believed that it really moved.

Despite its mathematical advantages Copernicus' model didn't really catch on for several decades and part of the reason for this may have been that, although they gave a natural explanation for the retrograde motion of the planets, Copernicus' circular, Sun-centred orbits still could not reproduce all of the observed details of planetary motion. In particular, they failed to account for the way the planets appeared to speed up and slow down as they moved against the background stars. Like Ptolemy before him, Copernicus had to resort to adding epicyclic motions to his model in order to explain this, leaving it open to the same charge of arbitrary complexity that had been levelled at geocentrism in the first place.

The next breakthrough was made by the German astronomer Johannes Kepler, using decades of detailed planetary observations that had been made by his mentor and former boss, Tycho Brahe. Brahe himself had been an interesting character. After losing his nose in a sword fight with a relative, he wore an artificial replacement made of brass and at one point he also kept a tame elk in his castle, which allegedly died after it drank too much beer and fell down the stairs. Brahe's own death is said to have been due to complications from a burst bladder, sustained during a royal banquet because he was too polite to leave the table in order to relieve himself. Despite these eccentricities, Brahe was a meticulous astronomer and his observations of stellar positions and planetary motions were among the most precise ever made. Well aware of Copernicus' work, he saw the mathematical sense in the argument that the Sun was the centre of planetary motion but he was unable to give up the idea of an immovable Earth at the heart of the cosmos. Instead, he championed a hybrid 'geoheliocentric' model, in which the Moon and Sun orbited the Earth while all of the other planets went round the Sun.

Kepler perhaps had a less colourful lifestyle than Brahe, but intellectually he was certainly more radical. He had embraced pure heliocentrism early in his career and was convinced that it could be reconciled with biblical scripture, with the central Sun representing God, the fixed stars representing Jesus and everything in between representing the Holy Spirit. If this sounded unorthodox from a religious point of view, his astronomical ideas were equally shocking. Despite disagreeing on whether the Earth or the Sun was at the centre of the cosmos, both Ptolemy and Copernicus had assumed that all celestial motions were based on the pleasing perfection of the circle. But, after studying Brahe's extensive data on the actual motions of the planets, Kepler concluded that they were moving not on circular paths but on elliptical ones, speeding up when they approached the Sun and slowing down when they moved further away. These elliptical orbits immediately did away with the need for Copernicus' epicycles, making the heliocentric model more streamlined and at the same time a better fit to what was actually observed in the heavens. Based on this model, Kepler formulated his famous three laws of planetary motion – and they are still taught today.

As a Protestant, Kepler was already irredeemably tainted with the stain of heresy as far as the Roman Catholic Church was concerned, but his ideas were not necessarily popular with his fellow Protestants either. In any case, whatever their revolutionary implications, Kepler's concepts were expressed in the language of mathematics and this made them inaccessible to much of the population, for whom the most compelling arguments came not from philosophical musing or mathematical theory but from the evidence of their own eyes.

It's astonishing to think that, up until the seventeenth century, all of astronomy's important observations and great discoveries had been carried out with the human eye alone. Even the painstaking observational work of Tycho Brahe had all been carried out with the naked eye. Aided only by devices for pointing and sighting, his measurements had achieved astonishing precision but, like all astronomical observations up until that point, they were ultimately limited by the constraints and frailties of the human visual system. This all changed in 1608, when news of a remarkable invention began to spread around Europe. The 'spyglass' was allegedly able to make distant objects appear to be much closer to the viewer, allowing more detail to be seen, and its military potential as a tool for spying on enemies ensured that rulers and generals across the continent received the news of its abilities with great interest.

The optical properties of curved pieces of glass had been known for centuries: Arabic scientists had honed the use of convex lenses to bend and focus light during the Middle Ages, and since the thirteenth century spectacles had been worn in Europe to correct poor eyesight. What was new about the spyglass was the combination of two different lenses to create a magnified image of a far-away object, allowing the viewer to resolve detail that would otherwise be too small for the unaided eye to make out. This ability is made clear in the name by which the device is more commonly known today: the telescope, from the Greek *teleskopos*, or 'far-seeing'.

The telescope's origins in the Low Countries in 1608 are shrouded in confusion. There are at least three plausible claimants for the title of inventor, but so straightforward was the principle behind the instrument that the authorities in The Hague refused to issue a patent to any of them. Simply reading a description of the telescope and its abilities was enough to enable craftsmen across Europe to reproduce the instrument for themselves – so this was an invention that would be very hard to control or suppress. Within weeks, news of the device had spread between centres of political and religious power via diplomatic dispatches: to Brussels, Paris, Madrid, London and swiftly on to the Venetian Republic where, in the summer of 1609, it reached the ears of a professor of mathematics named Galileo Galilei.

Galileo swiftly constructed his own telescope, and soon after he also modified the design to give much greater magnification. But as well as improving the instrument's performance he was also among the first to recognize its full potential, not simply as a tool for gaining military advantages here on Earth but as a means of studying the heavens in more detail than had ever been possible before. By turning his telescope towards the sky, Galileo was able for the first time in history to see things that were too small and far away for the eye to perceive. It was as if a new window had suddenly opened onto the universe and, in a feverish burst of activity in 1609 and 1610, Galileo made a series of revolutionary discoveries, each of which served to fatally undermine the traditional geocentric view of a cosmos in which everything revolved around the Earth.

Through his telescope, Venus, the brightest of the planets, no longer appeared as a dazzling point of light but as a round disc which, like the Moon, could be seen to undergo a series of phases, from crescent, to gibbous, to full, over a period of several months. It was fairly straightforward to show that this pattern of illumination could only be explained if Venus was circling around the Sun, rather than the Earth.

The Moon itself also revealed its secrets to Galileo's telescopic gaze. Long regarded as a smooth globe of perfect, incorruptible heavenly stuff, the Moon's prominent light and dark markings had been dismissed as the result of regions of different density rather than indications of texture or roughness. Through the telescope, a very different picture emerged: the Moon was not a smooth, ethereal realm. Instead, there were rugged mountains and deep craters casting shadows that lengthened and shortened in the raking sunlight as the Moon went through its cycle of phases. Like the Earth, this was a world with its own landscapes. And if the Moon, which moved through space, was also a world like the Earth then was it not possible that the Earth was like the Moon and therefore might also move?

Perhaps the most significant of Galileo's discoveries came when he turned his telescope towards Jupiter. Like Venus, this bright, steady point of light was resolved into a small disc, but immediately apparent through the telescope were several accompanying dots of light on either side of Jupiter, all arranged neatly in a line. These dots were invisible to the naked eye, and no one had ever suspected their existence, but as Galileo returned his attention to Jupiter on subsequent nights an even greater surprise was in store. From night to night, the small dots – four of them in total – changed their positions relative to Jupiter, always staying in a line with the planet but moving from one side of it to the other and sometimes even disappearing behind it. The only plausible explanation for this new phenomenon was that Jupiter's four accompanying points of light were moons – satellites of the planet that moved around it, not around the Earth or even the Sun.

The implications of these observations were not lost on Galileo, and it is perhaps a testament to the excitement of those few days in early 1610 that the first record of Jupiter's moons is a diagram found scribbled on the back of a used envelope. No longer could anyone claim that everything in the sky appeared to move around the Earth. Moreover, if Jupiter – a wandering planet – had satellites circling around it just as Earth did, then surely it was possible that the Earth – like Jupiter – might itself be in motion.

Ever the opportunist, Galileo rushed his discoveries into print as the *Sidereus Nuncius* or 'Starry Messenger', taking care to curry favour with the powerful Medici family by referring to Jupiter's satellites as 'the Medicean Stars' in the hope that they would offer him a job in Florence. His message was clear: through the telescope, it could be seen that the old, geocentric model, in which a perfect celestial realm revolved about the Earth, simply could not be true. A heliocentric picture was far more in keeping with this new, telescopic sky. The book was quickly distributed across Europe and caused a sensation – the English ambassador to Venice even described it as 'the strangest piece of news ... ever yet received from any part of the world'. Like many scientists of the time, Galileo was an accomplished artist and the beauty and explanatory power of his illustrations to the *Sidereus Nuncius* helped to ensure that its message was widely understood. The Medici were also impressed, and Galileo was able to take up a prestigious (and better-paid) new role under their patronage.

The reaction to Galileo's discoveries was not as negative or as one-sided as we might expect. Like many people, the English poet and clergyman John Donne was concerned that, by overturning the familiar hierarchy of the heavens, a challenge had been issued to the established social order here on Earth. Others feared that the Jovian satellites would throw astrology, the calendar and even medicine (at the time, heavily influenced by astrological analogies) into confusion. But the brand-new sky revealed by Galileo's telescope also provoked wonder and excitement, even among high-ranking members of the Church. Official acceptance of the heliocentric model might be a step too far, but Galileo's revelation of an imperfect, cratered Moon chimed neatly with its traditional role in Catholic iconography as a symbol of inconstancy – and within months the artist Ludovico Cardi (known as Cigoli) had incorporated a rugged lunar profile into his fresco of The Assumption of the Virgin in the Pauline

Chapel of the Basilica Santa Maria Maggiore in Rome, with the approval of none other than the Pope himself.

This just goes to show that the popular version of Galileo's history – that of a lone scientific pioneer bravely speaking out against the intellectual oppression of the Church – is not the whole story. He may have been a highly talented scientist and mathematician, but Galileo was no politician. His ideas initially had many advocates at the Vatican and, in 1616 when the Inquisition announced its first verdict on the topic, Galileo was not completely banned from mentioning heliocentrism: instead he was allowed to discuss it as a purely mathematical construct but not to advocate it as the true system on which the cosmos was based. However, as well as allies Galileo had made several powerful enemies in Rome. Over the next few years, through a combination of rash acts and incautious remarks, he proceeded not only to antagonize them further but also to enable them to build up a catalogue of evidence against him.

In 1632, with permission from both the Pope and the Inquisition, Galileo returned to the subject of heliocentrism versus geocentrism in his *Dialogue Concerning the Two Chief World Systems*. As the title suggests, the book presents the opposing arguments in the form of a conversation between two fictional characters – a clever stratagem that should have allowed Galileo to avoid the appearance of taking sides. However, in what now seems a catastrophic misjudgement, Galileo left the careful reader in no doubt of which position he really favoured. The character advocating the old, Earth-centred model was named 'Simplicio', which Galileo claimed in his introduction was a reference to

Simplicius, a respected classical philosopher. Of course, as his readers would have been well aware, in Italian *simplicio* can also mean 'simpleton' and, sure enough, in the book the character is prone to embarrassing lapses of logic. Worse still, Galileo placed into the hapless Simplicio's mouth pronouncements that had originally been made by the Pope himself.

This was the blunder that Galileo's enemies had been waiting for. Not only had he disobeyed the injunction of 1616, banning him from advocating heliocentrism, but he had also insulted the head of the Church. A second trial was swiftly called, and this time the verdict was harsh and unambiguous. In 1633, Galileo was found 'vehemently suspect of heresy' and placed under house arrest – a situation that he endured until his death in 1642. His legendary status as a scientific martyr was assured, but perhaps if he had played his cards more astutely the outcome could have been very different. It is fascinating to imagine how the subsequent history of science might have played out if the Church had been persuaded to champion, rather than condemn, the idea that the Earth is a planet moving around the Sun.

Either way, it is far too simplistic to present the story of heliocentrism as one of science versus religion. Copernicus, Brahe, Kepler and Galileo were all devout believers as well as men of science, and for them the increasing discrepancy between traditional teachings and what they observed in the sky was more likely to have been a source of discomfort than triumph. Instead, their hope was that religion, properly interpreted, and science, properly conducted, would converge on a new and better way of understanding the cosmos and the place of human beings within it. Meanwhile, inside the Catholic Church and in other religious organizations across Europe, there were many educated people who were fascinated and convinced by the new observations and ideas. But, as is often the case, politics, power and vested interests were the factors that decided Galileo's fate.

Ultimately, however, the evidence of the telescope could not be denied. The Roman Church's ban on books by Copernicus, Kepler and Galileo was impossible to apply rigorously even in Catholic countries. Besides, anyone with access to a spyglass could see for themselves the telltale phases of Venus, the craters of the Moon and Jupiter's orbiting satellites and so, as the seventeenth century wore on, heliocentrism inexorably gained ground. In 1687, the English mathematician Isaac Newton published his *Philosophiæ Naturalis Principia Mathematica*, or Mathematical Principles of Natural Philosophy, in which he set out his law of universal gravitation, at last providing a theoretical underpinning for the heliocentric model. Newton's law showed how the same force of gravity that caused an apple to fall from a tree here on Earth was also responsible for holding planets in orbit around the Sun, and it even provided a natural explanation for Kepler's observation that the planets speeded up and slowed down as they moved around their elliptical orbits. An account of the mechanism that held the cosmos together was a welcome adjunct to the heliocentric model. Elliptical orbits were incompatible with the idea of solid spheres holding the planets in place, while the discovery that comets actually crossed these planetary orbits meant that, whatever supported them, it certainly wasn't rigid and impenetrable crystal.

The wheels of the Church moved more slowly but in 1741 Pope Benedict XIV allowed a mildly censored version of Galileo's complete works to be published and the Vatican dropped its ban on the teaching of heliocentrism in 1835. Pope Pius XII, speaking in 1939, described Galileo as one of the 'most audacious heroes of research' and in 1992 Pope John Paul II officially expressed regret for the way that the astronomer had been treated almost four centuries earlier. But, already, by the end of the seventeenth century, few educated people seriously doubted the idea that, like Mercury, Venus, Mars, Jupiter and Saturn, the Earth was a 'wanderer', a planet in orbit around the Sun. In 1692, Newton's colleague, the astronomer Edmond Halley, was able to state without public controversy that, 'It is now taken for granted that the Earth is one of the Planets.' The heliocentric model provided a simple, elegant explanation for the observed motions of everything in the sky as the combined effect of living on a spinning planet, orbited by the Moon, and moving around the Sun along with all of the other planets. But, for all the persuasive mathematics of Copernicus, Kepler and Newton, or the observational skills of Brahe and Galileo, there was still one missing piece of evidence: physical proof that the Earth is in motion.

This was not necessarily surprising. Galileo's experiments with moving objects had already led him to formulate ideas about inertia, or the resistance of a body to changes in its motion, and these were further refined by Newton later in the seventeenth century. Newton's first law of motion, published alongside his theory of gravitation, states that a moving object will continue to travel in a straight line unless it is acted upon by an external force. As anyone who has travelled in a high-speed train will know, it is changes in motion rather than motion itself that we feel. On a straight section of track, a passenger on the train would need to look out of the window in order to tell that they were in motion, but when the track curves, forcing the train to deviate from its straight path, we feel the change in motion as our seat pushes against us, forcing us to follow a curved path too.

Of course, people standing on the surface of the Earth are not moving in a straight line either: we are spinning about the axis of the Earth as well as travelling around the Sun along an elliptical path, so the direction of our motion is constantly changing. In both these cases, gravity is the force that keeps us on these curved trajectories, preventing us from being hurled from the surface of the planet and stopping the Earth itself from flying off into space in a straight line. However, the Earth is very large and its orbital path around the Sun is even larger: unlike a train taking a tight corner, the curvature of our rotational and orbital paths is rather shallow and the change in our motion on a moment-by-moment basis is therefore too small for our senses to register. The effects of our constantly curving movement are now measurable by modern instruments and, in fact, we now know that the spin of the Earth plays a considerable role in shaping our planet's climate via the inertia of the atmosphere and oceans – something that we will explore in the next chapter (Spinning Around). But in the seventeenth and eighteenth centuries, scientists looked to other ways of verifying the Earth's motion through space.

One attractive candidate was the idea of stellar parallax – the expected changes in the appearance of the constellations as the Earth moves from one side of the Sun to the other. The Ancient Greeks had looked for this effect and failed to find it, but this might simply be because the stars are very far away and the annual changes are therefore correspondingly small. Perhaps now, with the aid of the telescope, these telltale stellar parallaxes could finally be detected, demonstrating that the Earth's position relative to the stars was changing throughout the year?

This was the goal of James Bradley, an astronomer from Gloucestershire who set out to detect stellar parallax in the early eighteenth century. In 1728, using state-of-the-art telescopic equipment – a 'zenith sector', which was carefully calibrated to maximize the precision of its measurements, and mounted to point straight up towards the zenith of the sky to minimize the refraction of the atmosphere – Bradley set out to monitor the apparent position of the star Gamma Draconis throughout the year. He was looking for a cyclical variation in its coordinates that would be consistent with the changing line of sight from Earth as our planet moved around the Sun.

After painstaking observations, Bradley detected a cyclical change but, to his surprise, it was not the change he was looking for. The position of Gamma Draconis varied by a tiny amount, but in the wrong direction and with the wrong timings to be due to parallax. After much thought, Bradley realized that he had detected a quite different phenomenon: one that we now call the aberration of light.

The principle of aberration is familiar to anyone who has ever used an umbrella. If you are standing still and the rain is falling vertically, then you need to hold your umbrella straight up in order to keep the rain off. But if you start to walk forwards, the vertical rain now appears to be approaching you at an angle, from the direction in which you are travelling, and to avoid getting wet you need to angle your umbrella slightly forwards to block the raindrops. Of course, the rain is still falling vertically – it is you that is moving – but the apparent direction from which the rain is falling now appears to be slightly in front of you rather than directly overhead, and the faster you walk the greater the apparent deviation of the rain from the vertical.

We see the stars because our eyes detect the light that they radiate towards us, and the direction from which its light appears to arrive at the Earth is what determines each star's apparent position in the sky. From our point of view, incoming light rays are a bit like raindrops: if we are moving through space, then the angle at which they appear to arrive will be shifted slightly towards the direction of our own motion – and this in turn will alter the star's apparent position. Since the orientation of the Earth's motion is constantly changing as it orbits the Sun, the amount and direction of aberration in the star's position will also change in a cyclical fashion throughout the year. Crucially, this annual variation should take exactly the form that Bradley had measured.

A full explanation of Bradley's observation requires the physics of Einstein's relativity, which was still almost 200 years in the future. But even with the classical Newtonian physics of his time, Bradley was able to show that the measured annual change in the position of Gamma Draconis could only be due to the motion of the Earth about the Sun. Although the effect was tiny – a wobble of less than a sixtieth of a degree – this was a momentous discovery. Bradley had set out to measure stellar parallax – which would have demonstrated that the Earth's position relative to the stars had changed over a six-month period, implying that our world must have moved through space between the observations. But the aberration of light was an even more direct demonstration of movement, since it showed that the Earth, with Bradley's telescope attached to it, was moving through space while the observations were being made. Although not the evidence that Bradley had set out to find, this evidence was just as good, if not better. As Galileo had asserted, the Earth is moving.

Bradley's measurements of aberration allowed him to refine the estimate of the speed of light, and even his failure to detect stellar parallax still had enormous scientific value since it demonstrated that Gamma Draconis – and by extension all of the other stars – must be much further away than anyone had previously believed, giving astronomers a new insight into the truly immense scale of interstellar space. Further observations with the same zenith sector telescope revealed another surprise: as the Earth spins on its axis our planet periodically sways from side to side, like the nodding motion of a spinning top. This swaying motion is called nutation and is due to the gravitational influence of the Moon, but it showed that the Earth's motion through space is anything but straightforward and its measurement was vital for refining the accuracy of navigational tables that used the stars and the Moon as guides.

Bradley's twin discoveries of the aberration of light and the nutation of the Earth's axis were the final nails in the coffin for the idea that our world is the unmoveable centre of the cosmos. Instead, they showed that the Earth is hurtling through space and wobbling on its axis as it goes. An argument dating back 2,000 years had finally been decisively settled and, although he is almost unheard of today, in his own time Bradley was feted across Europe. His discoveries were described as 'the most brilliant and useful of the century' by the French astronomer Jean Baptiste Joseph Delambre, and in due course he succeeded Edmond Halley as Astronomer Royal at Greenwich.

But this was far more than just the resolution of an academic debate. For centuries, the apparent motions of the sky had been seen as a reassuring demonstration of humanity's special, central place in the universe and, by tracking and describing those motions, astronomy simply provided additional evidence of our own cosmic significance – as well as lending support to the hierarchies of political and religious power that claimed to mirror the divine order of the heavens. But, by showing that it was the Earth itself that moved, Copernicus, Kepler, Galileo and Bradley had turned this relationship on its head. Instead of supporting the status quo, astronomy questioned and undermined it. Instead of appealing to accepted wisdom, astronomy looked to hard evidence and rigorous argument. And, unlike divine revelations and esoteric knowledge, reason and evidence were tools that were accessible to anyone – and so the authority to pronounce on the workings of the natural world could no longer be confined to a self-appointed elite. Along with developments in other sciences such as medicine, biology and chemistry, this change in approach set the scene for the eighteenth century's Age of Enlightenment and helped to shape the democratic, technological world that we inhabit today.

Bradley's zenith sector is now on display at the Royal Observatory in Greenwich just metres away from the telescope that defines the Bradley Meridian, the line that was used during his time as Astronomer Royal as the zero point for measuring both longitude and time according to the Sun. Although the Bradley Meridian was superseded as the official Greenwich Meridian in 1854 by a new line a few metres further to the east, it is still used as the reference point for longitude by the UK's Ordnance Survey maps, familiar to generations of British ramblers and hikers (having used Bradley's meridian since 1801, the Ordnance Survey were reluctant to redraw half a century's worth of maps on the whim of a new Astronomer Royal). Thus, the telescope that proved that the Earth is wandering through space is now side by side with the telescope used for regulating clocks and fixing positions down on the ground. This juxtaposition illustrates astronomy's contradictory potential: it can provide a comforting anchor with which to define and constrain the world around us, but at the same time its revelations can unseat our most cherished beliefs, casting us adrift in a vast and indifferent universe.

Despite his triumphs, Bradley never achieved his original goal of discovering stellar parallax. This would have to wait until 1838, when a century of improvements in telescope technology allowed Friedrich Bessel to detect tiny changes in the position of the star 61 Cygni over a six-month period, as the Earth moved from one side of the Sun to the other. The star's parallax was tiny – an angular change of less than a ten-thousandth of a degree – and required extreme precision to detect it, equivalent to being able to resolve an object 1 centimetre in diameter from a distance of over 3 kilometres away. Additional proof of the Earth's motion around the Sun was really rather superfluous by this stage, but the parallax of 61 Cygni was important for another reason: it allowed the distance to the star to be calculated accurately for the first time.

The principle was really very simple, relying on basic trigonometry, but the measurement of stellar distances from their parallax was only possible because the Earth changes its position throughout the year, allowing us to view 61 Cygni from two separate viewpoints separated by around 300 million kilometres – the diameter of the Earth's orbit around the Sun. Without this annual motion of our planet, it would not be possible for us to measure how far away the stars really are, and so our current knowledge of the vast scale of the universe depends on the fact that we are moving through space.

Bessel's measurement of the tiny amount of parallax demonstrated that 61 Cygni was extremely far away indeed – more than 100 trillion kilometres. As the parallaxes of other stars were gradually detected, their distances from the Sun were also revealed, and it became clear that many of them had distances even greater than that of 61 Cygni. The universe was clearly far larger than anything imagined by the Greeks, who had enclosed it within the crystal sphere of the fixed stars, situated at a comfortable distance just beyond Saturn.

Quantities like 100 trillion kilometres are extremely unwieldy to write (100,000,000,000,000), let alone use, and so astronomers were forced to abandon Earthly measures of scale such as miles and kilometres, and come up with a unit that was more appropriate for the task of representing interstellar distances. Travelling at over 300,000 kilometres per second, light travels around 10 trillion kilometres in one year. This figure – the distance travelled by a ray of light in one year or a 'light year' – is a suitable basis for attempting to understand the gulfs between the stars. It has become the standard yardstick for measuring distances outside the Solar System: 61 Cygni is 11.4 light years away from us, while the nearest star of all, Proxima Centauri, lies at a distance of 4.2 light years.

Such colossal distances might seem overwhelming, but they also give rise to another, even more mind-blowing implication: if it takes light 11.4 years to travel to us from 61 Cygni, then the light arriving right now from this star is actually showing us what it looked like 11.4 years ago, when the radiation first set out towards us. When we look into the night sky, we are peering back into the history of the cosmos, since we are literally seeing stars and other distant objects as they were in the past. To see them as they are right now we would have to wait for the light that they are currently emitting to travel to us. Telescopes are therefore the closest thing we have to time machines: not only do they make distant objects appear closer – the capability which had proved so useful to Galileo – but they also allow us to see deep into the past, and to watch events that occurred a long time ago as they unfold. In this sense, astronomy is a study of the past, like archaeology or history. The deeper we look into space, the earlier the time period we are observing: astronomy is a science of origins, which allows us to piece together the stories of how galaxies, stars and planets formed and evolved by directly observing these ancient processes as they happen. Bound up in all of this distant, primordial activity is the story of our own origins – and so the study of the remote cosmos is ultimately also the study of us.

The revelation of the vast scale of space is yet another example of the way in which astronomical discoveries continue to make us seem ever smaller and less significant – a process that started with the rejection of Ptolemy's geocentric model of the universe and the relegation of Earth to the status of a wanderer. Copernicus had argued that the Earth does not occupy a privileged, central position in the universe, but is instead just one planet among several others all orbiting the Sun. This assertion that we are not at the centre of things is now known as the Copernican Principle, and it has come back to haunt us several times over the last 500 years, as new discoveries have repeatedly displaced us further and further from any idea of a cosmic centre – and even challenged the idea of there being a centre at all.

When Copernicus set the world in motion, he at least allowed people to cling to the idea that the Sun was at the centre of the cosmos, with everything else, including the stars, arranged around it. Earth might not be the stationary point around which everything else moved but human vanity could take comfort from the thought that at least we weren't all that far from the cosmic hub itself, in the form of the Sun – not the prime position perhaps, but still a fairly significant one. But, once it had begun, the Copernican erosion of the Earth's importance proved unstoppable and, with each subsequent discovery, the human ego has taken another battering. Astronomy has repeatedly revealed a universe that is even larger than we had previously imagined and – within that vastness – an Earth that is further than ever from any kind of central or unique position.

Bessel's measurement of the distance to 61 Cygni was a major breakthrough because it allowed astronomers to begin to accurately map the three-dimensional distribution of the stars throughout the cosmos. But by 1838 it was already abundantly clear that the stars were not spread evenly through space and this had disturbing implications for the idea that the Sun was in a special position at the centre of creation.

The Copernican, heliocentric model of the universe had dispensed with the need for crystal spheres to hold the planets in place. And if the planets floated freely, then perhaps there was no need for the stars to be fixed to a sphere either – could they not instead be scattered through space at many different distances? In 1584, the Dominican friar, philosopher and mathematician Giordano Bruno went even further, suggesting that each star was in fact a sun with its own solar system of planets orbiting around it and that some of these planets might even support life. He also made the shocking suggestion that the universe might be infinite, without a centre or edge, and that the position of the

Sun was therefore no more important or unique than that of any other star. As prescient as his ideas seem to us now, Bruno was not a diplomatic man and – like his contemporary Galileo – he managed to make some powerful enemies both within the Church and in academia. Bruno also espoused unorthodox views about several articles of Catholic faith, such as the Virgin Birth and the Holy Trinity. As Bruno was a member of the Church himself, the Roman Inquisition took a particularly dim view of all of this. Bruno was found guilty of heresy and burned at the stake in 1600.

Rather ironically, as the heliocentric model caught on during the seventeenth century, so too did the idea that the stars were suns in their own right and were scattered through space at different distances. But if each star is a sun, then this also means that our own Sun is just another star – one of thousands visible to the naked eye and millions more visible through a telescope. Thus, the ultimate consequence of the heliocentric worldview was not just to displace the Earth from the centre of the cosmos and replace it with the Sun, it also removed the Sun itself from any kind of privileged, central position. The heliocentric model had contradicted its own name: the Sun might be the centre of the Solar System, but it was not the centre of the universe.

This posed an uncomfortable question. Newton's law of universal gravitation had successfully explained how the Sun held the planets in orbit around it, by providing an inward force to keep them on closed, elliptical paths and counter their natural tendency to keep moving in a straight line. Gravitation and motion are in balance in the Solar System: without the Sun's gravity, the planets would fly off into the depths of space, but without their motion they would quickly be pulled into the Sun and be destroyed.

However, Newton's theory required gravity to be a universal force and if it acts to pull an apple towards the Earth and to pull planets towards the Sun then it should also act to pull the Sun and all the other stars towards each other too. A universe of stars scattered randomly throughout space should be inherently unstable: gravity would inevitably yank them from their places and ultimately cause the universe to start collapsing about our ears. Clearly this was not the case, so there must be a missing ingredient to the picture. Newton himself proposed a solution: if the stars were distributed perfectly evenly throughout an infinite universe, then each star should be attracted equally in all directions by the combined gravitational pull of all its neighbours. In this perfectly balanced scenario, the universe could remain static and unchanging – but only if nothing happened to disturb the even distribution of the stars. Newton was aware that the slightest disruption could throw the whole system out of balance: a slight concentration of stars in one region of space would start to pull other stars towards it, increasing the imbalance and pulling in yet more stars in an unstoppable chain reaction.

Solutions that require such delicate fine-tuning have a tendency to be unconvincing – and they beg the obvious question of how the system arrived at its state of perfect balance in the first place. Besides, a quick glance at the night sky is enough to suggest that the distribution of stars is far from even. Through a telescope, the unevenness is even more pronounced: another of Galileo's discoveries had been that the Milky Way – the band of cloudy light that encircles the entire sky – was in fact made up of millions of individual stars, demonstrating that there were far more stars in some directions than in others.

Towards the end of the eighteenth century, the astronomer William Herschel set out to explore the full extent of the problem. Without being able to measure the distance to any of the stars – Bessel's measurement of parallax was still decades in the future – he carried out detailed star counts in 600 small patches of sky distributed around the heavens, making the reasonable assumption that the more stars he could see in a particular patch of sky, the further the distribution of stars must extend in that direction. The work was painstaking and involved many cold nights glued to the eyepiece of a telescope, meticulously noting every star in the field of view, then moving on to the next. When Herschel's survey was complete and the data was all collated, it was very clear that the star counts were not constant all over the sky. From this, he was able to piece together a rough picture of the way the stars were distributed throughout space, and of the position of the Sun within that distribution.

Herschel likened the stellar distribution to a giant millstone – a flat, disc-shaped structure that was much wider than it was thick. The Sun lies inside the disc itself and so, from our point of view, when we look through the plane of the disc we see many more stars than we when we look 'up' or 'down' out of the disc. This gives us the impression of a dense band of stars encircling the sky: our familiar view of the Milky Way. But even the band of the Milky Way is not evenly blessed with stars. Herschel's star counts showed a significant increase towards the part of the Milky Way that ran through the constellation Sagittarius and he reasoned that the stars extended much further from us in this direction than it did in others. In other words, although the Sun lies within the disc of the Milky Way. it is certainly not at the centre. Once again, the Copernican Principle had made itself felt. Not only is the Sun just one star among millions, it can't even claim to occupy a special place among its stellar neighbours. Our understanding of the distribution of the stars had also shifted, and the Milky Way had gone from being simply a band of pale light encircling the sky – *galaxías kýklos* or 'milky circle' in Ancient Greek – to a disc-like structure encompassing all of the visible stars including the Sun – in other words, a galaxy.

Herschel's ability to map the Milky Way was constrained by the observational limits of eighteenth-century telescopes and by the time-consuming requirements of making detailed star counts in as many directions as possible. Unknown to astronomers of the time, a further constraint came from the fact that the Milky Way contains vast clouds of interstellar dust, which absorb the light of the stars behind them and effectively limit the visible part of the Milky Way to a small fraction of its total extent (*see* Made of Starstuff, page 6). Nevertheless, Herschel's 'millstone' was very much along the right lines and two centuries of careful observation and interpretation, along with huge advances in telescope technology, have gradually revealed the full structure and extent of the Milky Way galaxy, at last allowing us to appreciate the nature of our local stellar environment – and the true place of the Sun within it.

The Milky Way galaxy is a disc of stars, dust and gas around 100,000 light years in diameter but only around 2,000 light years thick. The disc swells in the middle into a crowded region known as the 'bulge', containing billions of ancient stars and a supermassive black hole with a mass equivalent to 4 million Suns, while several spiral arms – dense lanes of stars and dust – swirl out from the centre towards the rim of the disc. The Sun is just one of between 200 and 400 billion stars in the Milky Way and it is situated in the disc around 27,000 light years from the galactic centre, on the edge of the Orion Arm. If the Milky Way is a city of stars, then the Sun is located somewhere in the outer suburbs.

Clearly, the stars are not evenly distributed throughout space and so Newton's suggestion that the stars are held in their places by an infinite array of balancing gravitational forces from every direction can't be correct. Instead, the reason that the Milky Way does not collapse to a point under its own gravity is that all of its constituent parts – gas, dust, stars and the Sun itself – are in constant motion. Just as the planets of the Solar System are in orbit around the Sun, everything in the Milky Way is orbiting around its centre of mass, which lies at the same position as the supermassive black hole in the middle of the galaxy. The reason we are here at all, rather than crushed to a point inside a super-massive black hole, is that we are moving.

As we have seen, the ancients referred to 'the constellations of the fixed stars' and regarded them as an unchanging backdrop against which the Sun, Moon, planets and other celestial objects made their journeys across the sky. We now know that the stars are far from fixed: on the contrary, every single star in the sky is moving through space, orbiting around the centre of the galaxy. Modern observing techniques even allow us to measure this motion directly either by looking for characteristic changes in the colour of light emitted by the star as it moves towards or away from us (an effect known as the Doppler shift, which is also responsible for the noticeable change in the pitch of a police car siren as it rushes past you) or by detecting tiny year-on-year changes in the positions of the stars – an ability that would have astonished Bradley or Bessel.

The Sun is no exception, moving through the disc of the galaxy at a speed of around 250 kilometres per second and making one complete circuit around the Milky Way's centre of mass every 220 million years or so. As well as looping around the centre of the galaxy, the Sun also oscillates up and down within the disc with an average speed of 7 kilometres per second, so its path through the Milky Way is an undulating one. Of course, while the Sun is moving through the galaxy, the Earth is also orbiting around it with a speed of around 30 kilometres per second, so our planet's galactic path is actually a spiral, coiling around the trajectory of the Sun.

The stars in our neighbourhood are moving in roughly the same direction but each follows its own individual path around the galaxy, and at its own speed. Like a swarm of bees, they are constantly changing their positions relative to one another, so the familiar stars in night sky are really only temporary companions on our long journey around the Milky Way and the patterns that they form in the heavens today are anything but eternal.

Over thousands of years, their motions with respect to us and to each other become more and more apparent. As our ancestors painted their Ice Age prey on the cave walls of Lascaux, the constellations that hung above them would have had noticeably different forms from those of today – and when the glaciers return in around 50,000 years' time our current constellations will have altered in their turn. Over the course of one entire orbit of the Milky Way – a 'galactic year' – the changes in the sky will be even more radical, as some stars speed on ahead of us while others fall behind. It is unlikely that a single one of today's stars would have graced the skies that hung above the dinosaurs.

Being in the middle of this storm of movement makes it difficult to see what's really going on in the Milky Way, so quite a lot of what we know about stellar motion actually comes from studying other galaxies where, although they are often millions of light years away, we have the advantage of seeing them from the outside.

DARK MYSTERIES

Global views of galactic motions have also helped us to understand something fundamental about the nature of the universe itself. Since the 1930s, astronomers had been finding disturbing hints that something was missing from our picture of what a galaxy is made of.

The situation came to a head in the 1970s when American astronomer Vera Rubin measured the orbital speeds of stars in a sample of spiral galaxies and compared them with the amount of gravity provided by all the stars, gas and dust that the galaxies contained. Something odd emerged from her results. The stars at the edges of the galaxies were moving surprisingly fast – so fast that a great deal of gravity was required to prevent them from flying off into space. But when Rubin added up all of the visible matter in the galaxies, there was simply not enough of it to account for this powerful gravitational field – and yet the high-speed stars remained in their orbits. Clearly something was missing: the galaxies had to contain substantial amounts of additional matter, but matter that was completely invisible to our telescopes. Rubin's work proved highly controversial – it was hard for the astronomical community to accept that a huge fraction of the matter in the universe had gone unnoticed for so long. But other studies soon backed up Rubin's work. It seemed that galaxies really did contain large amounts of invisible matter, whose presence could only be inferred by its gravitational influence.

There are now several independent lines of evidence that all point towards this same conclusion: the way that the gravity of galaxy clusters bends the light from more distant objects, the behaviour of hot gas in and around galaxies, and tiny fluctuations in the radiation of the Cosmic Microwave Background all indicate the presence of large quantities of matter that neither emits, reflects nor absorbs light. The name given to this mysterious stuff, 'dark matter', reflects this lack of light, as well as its unknown nature and its reluctance to interact with electromagnetic radiation.

Four decades on, dark matter remains an elusive quantity – we have yet to isolate a single particle of it and we still have no idea what it's actually made of. However, during the 1990s astronomers carefully observed its gravitational influence on the neighbouring, visible matter, and made huge strides in pinning down exactly how much of it there is and how it is distributed throughout the universe. Dark matter outweighs ordinary matter – the atoms that make up stars, planets and us – by a factor of more than 5 to 1. Unlike ordinary matter, which clumps together into dense accumulations such as planets and stars, dark matter particles seem to remain aloof even from their fellows. Instead, the dark matter forms huge, diffuse clouds, in which the galaxies of visible matter are embedded like the currants in a cake.

Over the 13.8-billion-year history of the universe, dark matter has exerted a profound influence on the way ordinary matter has behaved, helping to shape the galaxies themselves and, as we shall see in a later chapter (Written in the Stars, page 308), it will play an important role in determining the ultimate fate of the universe.

When the dark matter results were applied to our own galaxy, they made sense of the stellar motions here too. It is fair to say that the orbital properties of the Sun are determined to a large extent by the gravitational influence of the Milky Way's dark matter – and yet we only know that this mysterious stuff exists because the stars are moving.

Constant motion is the natural state of affairs within the Solar System and within the Milky Way and the other galaxies, so it should come as no surprise that the galaxies themselves are also moving. Our immediate galactic environment is a throng of over 50 other galaxies known as the Local Group, which is spread throughout a region of space roughly 10 million light years across. Like the stars within the Milky Way, each galaxy of the Local Group is moving along its own individual trajectory but the Group itself is 'gravitationally bound' because none of the galaxies is moving fast enough to be able to break free from the combined gravitational pull of its neighbours.

Unlike the Solar System or the Milky Way galaxy, the Local Group has no well-defined centre around which all its galaxies are moving. However, dominating the gravitational landscape of the Local Group are its two largest galaxies: our own Milky Way and the even larger Andromeda Galaxy – otherwise known as Messier 31. These two heavyweights rule the Group, and most of the smaller galaxies are effectively in orbit around one or other of them – in some cases they are even being pulled apart and cannibalized by their larger neighbours. But Andromeda and the Milky Way also work their gravitational influences on each other. Currently, the two are separated by around 2.5 million light years, but recent studies by the Hubble Space Telescope have confirmed that they are getting closer together, rushing towards each other with a combined speed of 110 kilometres per second. At this rate it will take around 4 billion years for Andromeda and the Milky Way to smash together in a spectacular galactic collision. We will explore the likely consequences that this will have for the Earth in a later chapter (Written in the Stars, page 308).

By now, it will come as no surprise that the Local Group as a whole is also moving through space. As groups of galaxies go, it is really rather a modest one and it is utterly dwarfed by its neighbour, a collection of over 1,000 galaxies called the Virgo Cluster, which spans a region of space around 100 million light years across. Like the members of our Local Group, Virgo's constituent galaxies are gravitationally bound within the cluster, executing a complex orbital ballet as they weave among their hundreds of neighbours. The combined gravity of this vast structure also exerts a powerful tug on the galaxies of the Local Group, and Virgo is pulling us towards it at several hundred kilometres per second – a motion called the Virgocentric Flow. But the Virgo Cluster doesn't have it all its own way and we aren't moving directly towards our mighty neighbour. There are several other large galaxy clusters in our corner of the universe and these are also exerting a gravitational pull on the Local Group, so that our overall motion is down to their combined effect, not just that of Virgo. Collectively, these giant clusters, along with the Local Group itself, are known as the Laniakea Supercluster. Laniakea is 520 million light years across and its borders encompass as many as 100,000 individual galaxies – appropriately its name comes from a Hawaiian word meaning 'immeasurable heaven'.

The overall motion of the Local Group is directed towards the centre of mass of Laniakea – a point known as the Great Attractor, and we are barrelling through space at a speed of around 600 kilometres per second. This may sound impressive but, even so, we are destined never to reach the Great Attractor, or even the less distant Virgo Cluster. They are forever out of bounds because, despite the fact that the Local Group is travelling in the right direction, the distance between them and us is getting larger all the time.

How can these galaxy clusters be getting further away from us even though we're travelling towards them? The most straightforward suggestion might be that they are rushing away from us faster than we are rushing towards them. But in fact this is not the case. They are not *moving* away from us – and yet the distance between them and us is still increasing. The fairly astounding explanation for this apparently paradoxical scenario was demonstrated by Edwin Hubble in 1929: the universe is expanding. Hubble's observations of this expansion confirmed an earlier prediction by the Belgian astronomer (and Catholic priest) Georges Lemaître, which was in turn derived from Einstein's general theory of relativity. In 1923, Hubble had proved that there were millions of other galaxies outside the Milky Way, greatly increasing the size of the known universe. But this newly discovered universe of galaxies posed a troubling question: what was stopping the mutual gravity of all these galaxies from causing them all to simply collapse together?

This was a similar problem to the one that had vexed Newton and his contemporaries in the seventeenth century: why are the Sun and the other stars of the Milky Way not being pulled together by their mutual gravity? Newton's original suggestion had been that the stars must be evenly distributed throughout space so that the combined pulls of stars from every direction worked to cancel each other out, leaving each of them at rest and perfectly balanced. But, as we have seen, the real answer is that the Sun and stars are all in constant motion orbiting around the centre of the Milky Way. Lemaître's solution to the conundrum of the galaxies was different from either of these scenarios. The galaxies and galaxy clusters are indeed pulling on each other gravitationally, but the reason they have not all collapsed together is that the space between them is constantly stretching out, increasing the distances between them so that, despite their mutual attraction, they can never reach each other. The idea seems to be profoundly at odds with our own daily experience of space as a fixed and unchanging quantity, but it makes perfect sense within the mathematical framework of Einstein's equations.

Counterintuitive or not, in 1929 Edwin Hubble was able to show that space behaved in exactly the way that Lemaître had predicted. The light from distant galaxies was being stretched out and reddened as it journeyed towards us through the expanding intergalactic space – a phenomenon now known as cosmological redshift. Hubble's key observation was that the further away a galaxy was, the more 'red-shifted' its light appeared – exactly as we'd expect, given that it had had to travel a larger distance in space to reach us and would therefore have been subjected to a correspondingly greater amount of stretching.

The only galaxies exempt from this rule are the Milky Way's closest neighbours, the 50 or so members of the Local Group. As we've seen, these galaxies are close enough together to be gravitationally bound to one another and their mutual attraction is therefore strong enough to overcome the general tendency of space to increase in volume. The same is true of other large collections of galaxies such as the Virgo Cluster: their mutual gravity is sufficient to hold them together against the expansion of space. But on a larger scale, these galaxy clusters are being inexorably separated from each other as the universe continues to expand, and vast groupings such as the Laniakea Supercluster are gradually being torn apart. From our point of view, all the billions of galaxies outside our own Local Group are effectively receding from us, as the distances between them and us grow ever larger.

This discovery was one of the key pieces of evidence that led to the Big Bang theory – the idea that the universe exploded from an incredibly hot, compact state around 13.8 billion years ago. If we turn back the cosmic clock, winding the expansion of the universe backwards, then space will contract and the galaxy clusters will get closer and closer together until everything merges into a single point – a singularity. As we've seen, the Big Bang model provides an impressively detailed account of how the universe got to be the way we see it today, from the distribution of galaxy clusters to the abundances of the chemical elements and the glow of Cosmic Microwave Background radiation that permeates all of space.

Hubble's observation that distant galaxies are all getting further away from us elegantly demonstrates the consequences of living in an expanding universe, but doesn't it also violate the Copernican Principle that there's nothing special about our place in the cosmos? After all, if the galaxies are receding from us the obvious implication is surely that our galaxy – or the Local Group at least – is situated at the very centre of the expansion. Could we be in the most privileged cosmic position of all, at the original site of the Big Bang?

Unfortunately for those who like to keep things straightforward, the answer is both yes and no. When we imagine the Big Bang and the subsequent expansion of the universe we quite naturally resort to comparisons with terrestrial explosions, in which matter and energy blast outwards from a central point and spread in all directions through a pre-existing, empty volume of space. But the Big Bang was not simply an explosion of matter and energy, it was the beginning of space and time as well. There was no pre-existing volume waiting to be filled and no unique coordinate at which the explosion occurred. In the instant of the Big Bang, space itself exploded into existence and the universe created the volume into which it expanded.

This weird scenario makes perfect mathematical sense but it is extremely hard for our brains to visualize it. Instead of everything rushing away from a single point, every point in the universe is expanding away from every other point. The galaxy clusters are effectively staying where they are, held together by their internal gravity, while the space between them relentlessly expands in all directions. From within each cluster, it will appear as though they are at the centre of this expansion while all of the other clusters are receding away from them – which is exactly what we observe from our vantage point here in our own cluster, the Local Group. As a rough analogy – although in two dimensions rather than three – we can imagine that the clusters are like spots painted on the surface of a rubber balloon. As the balloon is inflated the rubber between the spots expands. Pick any spot as your reference point and it will seem as though all of the others are receding from it, even though no single spot is in a more privileged part of the balloon's surface than any other.

To return to our own three-dimensional universe, every point in space is an expanded relic of the original point, the singularity at which the Big Bang occurred 13.8 billion years ago. The answer to the question 'Where did the Big Bang happen?' is 'Everywhere – but everywhere was all in the same place.' In a sense, we have come full circle: like the proponents of Ptolemy's geocentric universe, we can tell ourselves that we are at the centre of the cosmos, but we must accept that the centre is also everywhere else – and the Copernican Principle is upheld on the largest scale we can imagine.

In the 400 years since Galileo was put on trial for suspected heresy we have come a long way– not just in our understanding of the cosmos but also simply in terms of the sheer distance that we have travelled through space during those four centuries. Aristarchus, Copernicus, Kepler and Galileo have all been vindicated many times over, and no matter how much it may feel that the ground beneath our feet is stationary, we now know that we are moving all the time: speeding around the Sun at 30 kilometres per second, circling around the centre of the Milky Way at 225 kilometres per second, and plunging towards Andromeda at 110 kilometres per second, while the Local Group speeds on its ever-lengthening journey towards the Great Attractor at 600 kilometres per second. The fact of our motion has profoundly changed the way we think of ourselves and of our place in the universe, but it has also allowed us to probe and investigate our surroundings in ways that would have been impossible from the surface of a stationary Earth. Clinging to the surface of a whirling speck in a universe of constant motion may not be quite so flattering to our egos, but it's hard to imagine a more exciting ride.

Spinning Around

The Earth is engaged in a series of epic journeys – around the Sun, around the centre of the Milky Way, and through the vast emptiness of intergalactic space – but there's also another sense in which our planet is always moving. Like almost every other astronomical object in the universe the Earth is spinning on its axis, in this case making one complete turn every 24 hours. Along with our planet's other motions through space this spin is not something we can feel – because we are spinning with it – but its effects are nevertheless quite profound. Without its spin, our world would be a very different place indeed – and perhaps even an uninhabitable one.

The spin of the Earth is a relic from the formation of the Solar System itself, and we can think of it as a natural consequence of two of the most ubiquitous phenomena in the universe: gravity and motion. Any object moving through space will be affected by the gravity of neighbouring objects, causing its path to curve towards them and sometimes even making it loop back on itself, becoming a closed orbit. Curving and looping motion is there-fore the norm wherever objects are close enough together to feel each other's gravity, and we say that objects moving in this way possess 'angular momentum'.

This would have been true of every particle of gas and dust in the collapsing cloud from which the Sun and the Solar System formed 4.5 billion years ago, with each atom and grain moving on its own individual and largely random trajectory, curving through space in response to the influence of its neighbours. If we could add up the random motions of every particle in the cloud we would find that most of them cancelled each other out – but it would be very improbable for all these random motions to cancel out entirely. Summed over all of its particles the cloud itself would still be left with a non-zero amount of angular momentum – a residual twist or swirl of motion that character- izes the cloud as a whole.

As the cloud condensed and the gas and dust particles were squeezed more tightly together the material retained this overall twist, and as the individual particles began to collide and stick to one another, the cloud would have gradually flattened out into an immense rotating disc with the gas and dust all swirling in the same direction about the centre. This flattening out of a cloud of orbiting particles is a common phenomenon in astro- physics, and it's responsible for everything from the incredibly flat rings of Saturn to the disc-like profile of spiral galaxies. In our Solar System it is the reason why the Earth and all of the other planets orbit the Sun in the same direction and in roughly the same plane – this is a relic of the rotating disc from which the Sun and the planets formed and it's the plane that defines the circle of 12 constellations that we call the Zodiac.

As the Sun formed at the centre, grabbing 99.8 per cent of the available gas, it also inherited a large fraction of the angular momentum of the disc. The Sun now spins on its axis roughly once every 30 days (although, as we saw in the chapter Sunshine, this varies with solar latitude) in the same direction as the planets orbit. In the remaining disc of material a similar process repeated itself on a smaller scale so that, as local clumps of dust, gas and debris condensed into embryonic planets, each one inevitably inherited a spin in the same direction as it was orbiting. This is why if we could look down on the Solar System from above the Sun's north pole we would see everything spinning on its axis in an anticlockwise direction, with the planets also orbiting the Sun and various moons orbiting the planets, all in the same anticlockwise sense and all in roughly the same plane

Within our Solar System there are several exceptions but, as is often the case, on closer inspection they only serve to prove the rule. The planet Venus spins on its axis in the opposite direction to its orbital motion around the Sun, while Uranus rotates on its side, rolling around its orbit with its polar axis pointing towards the Sun for much of its year. It has been suggested that both these anomalous spins might be due to collisions with other planet-sized objects during the formation of the Solar System – similar to the impact that created the Earth's Moon – although in the case of Venus a tidal interaction between the Sun and the planet's dense atmosphere might also be responsible. Several moons, such as Saturn's Phoebe and Neptune's Triton, also orbit their parent planet in the opposite sense to its spin and its orbit around the Sun. It is thought that they were once free-floating objects that were gravitationally captured at some point in the past. As we learn more about the planetary systems around other stars we are beginning to see that, as expected, they follow a similar pattern with the majority of planets orbiting in the same direction, although some Hot Jupiter planets orbit their star the 'wrong' way, perhaps because they have been captured from other stars. As yet we can't see in which direction the majority of exoplanets spin on their axes but the expectation is that it will be in the same direction as their orbit, just as we find for planets in our own Solar System.

The most obvious effect of the Earth's rotation is the cycle of day and night and the apparent daily motions of the Sun, Moon, stars and planets around the sky. Each morning, the Sun appears to rise above the eastern horizon, crossing the sky throughout the day and setting in the west in the evening. The Moon, planets and stars also obey this east-to-west rule although, depending on your latitude, some of the stars closest to the north and south poles of the sky (the points directly above the north and south poles of the Earth) will never drop below the horizon – instead tracing endless circles around the pole. Just as scenery viewed from the window of a speeding train appears to fall behind as the train rushes forwards, this apparent east-to-west motion of the heavens is an illusion caused by the west-to-east motion of the spinning Earth.

TIME BY THE SUN

In fact, the Sun's daily circuit of our skies isn't entirely due to our planet spinning on its axis. Even if the Earth were not spinning at all, it would still be orbiting the Sun, and from our point of view this would make the Sun appear to move once around the sky for each complete orbit the Earth made around it – in other words, we would have a day that was the same length as our year. As well as being very long, this 'orbital day' would have another odd property: the Sun would appear to move the wrong way around the sky – rising in the west and setting in the east, in a complete reversal of what we are used to.

This means that the Earth's spin and the Earth's orbital motion are working against each other: the former makes the Sun appear to move westwards across the sky, while the latter is trying to move it eastwards. The Earth actually makes a complete turn on its axis every 23 hours and 56 minutes, but during this time it has also moved about 1/365th of the way around its orbit. Because there are 360 degrees in a circle, this means that the apparent position of the Sun is dragged back by an average of one degree every time the Earth makes a full turn – delaying the Sun's return to the same part of the sky by about 4 minutes. The combination of a 23-hour and 56-minute spin and a 4-minute orbital 'delay' is what gives us our average day length of 24 hours.

But the devil is in the detail and this really is only an average: in fact the exact contribution of the Earth's orbital motion varies appreciably throughout the year. The Earth's rotation axis is tilted relative to its orbit and it is this tilt that gives us our seasons, as the Earth spends half of the year with its northern hemisphere leaning towards the Sun and the other half with the southern hemisphere in the favoured position (*see* Inconstant Moon, page 153). From the ground, this causes the Sun's apparent position in the sky to move northwards for half the year – making it higher in the sky of the northern hemisphere during its summer – and then southwards for the other half – making it higher in the southern sky during its summer. The amount of north–south motion varies as the Earth moves around its orbit and this in turn affects the apparent speed of the Sun's east–west motion around the sky.

Another variation in the apparent motion of the Sun comes from the fact that our planet's orbit is an ellipse rather than a perfect circle, and the Earth moves along it at different speeds throughout the year depending on how close it is to the Sun. This means that the amount of delay in the Sun's apparent motion also varies throughout the year, with the fastest orbital speeds occurring around 3 January, when the Earth is at its closest to the Sun. (This is of course when the northern hemisphere is tilted away from warming solar rays, so perhaps we should be grateful that the severity of our northern winters is offset – if only in a very small way – by the planet's proximity to the Sun during this period. Six months later, the Earth is furthest from the Sun and orbiting at its slowest, so inhabitants of the southern hemisphere have their winters exacerbated – again only very slightly – by the increased distance.)

Between them, these two effects of axial tilt and varying orbital speed mean that the time taken for the Sun to make a complete circuit of the sky is exactly 24 hours on only four occasions during the year: around 15 April, 13 June, 1 September and 25 December. During the rest of the year, the length of a day – measured as the time taken for the Sun to return to its highest position in the sky – can be longer or shorter than 24 hours by up to 15 minutes or so. Only if we take the average, or mean, of all the days in a year does the day length come out as 24 hours. It is this averaged time that our clocks and calendars keep, and it is the origin of the 'mean' in Greenwich Mean Time – but it is rarely exactly in sync with the true position of the Sun in the sky.

In the days before super-accurate atomic clocks, the apparent motion of the Sun was used as the benchmark for timekeeping accuracy, and regular measurements of the Sun's position were the foundation of time systems such as Greenwich Mean Time (GMT). Each day at the Royal Observatory in Greenwich, astronomers would take a sighting of the Sun as it passed directly over the famous Greenwich Meridian. Since a meridian is simply a north–south line running from pole to pole, when the Sun crosses the meridian passing through Greenwich it is halfway through its daily journey from the eastern to the western horizon and is at its highest point in the sky for that day – in other words, it is noon in Greenwich. By noting the position of the Sun and comparing it with the time on their clocks, the astronomers could ensure that the clocks were running correctly but – because clocks stick rigidly to a 24-hour cycle and the Sun does not – the astronomers first had to convert their solar measurements from the Sun's 'apparent solar time' to the clocks' 'mean solar time'. Without the assistance of electronic computers, these calculations could take around an hour for the astronomers to carry out and this is why the famous red Greenwich Time Ball – mounted on the roof of the Observatory in 1833 so that ships on the River Thames could set their own clocks to GMT – is still dropped at 1pm every day rather than 12 noon.

Other planets of the Solar System also have elliptical orbits and tilted rotational axes and, in some cases, the resulting variation of day length as they move around the Sun is even more pronounced than that of the Earth. The orbit of Mars is significantly elliptical – far more so than Earth's – and the actual noon-to-noon length of a day on Mars can be up to 50 minutes longer or shorter than an averaged 'Martian Mean Time' clock would show. Meanwhile, the planet Uranus has an axial tilt of almost 90 degrees, making it appear to roll around its orbit and giving it a huge variation in day length as it moves around the Sun.

Although everything on Earth makes one whole revolution roughly every 24 hours, this doesn't mean that it is all moving eastwards at the same speed. Points on the equator have to travel a lot further to complete a revolution than points close to the poles. The circumference of the Earth's equator is 40,075 kilometres, so anyone standing on the equator makes a circular journey of this distance every 24 hours, and is therefore rushing towards the east at a dizzying 1,670 kilometres per hour. By contrast, a person standing 1 kilometre from the north or south pole only has to travel 6.3 kilometres to make a complete circuit – so they are moving at a much less hectic rate of 0.25 kilometres per hour. (Someone standing exactly at the pole itself will simply turn leisurely on the spot once per day.)

The rotational speed of objects on the Earth's surface can actually be put to good use by the space industry. Launching an object into space is hugely expensive: a rocket needs to have enough fuel not only to lift itself and its payload hundreds of kilometres above the Earth's surface – forcing its way through the atmosphere as it goes – but also to boost the payload to extremely high speeds in order to ensure that it remains safely in orbit rather than falling back to Earth. At altitudes of around 500 kilometres, known as 'low Earth orbit', a satellite needs to be travelling parallel to the ground at around 28,000 kilometres per hour – a staggering 7.8 kilometres per second, or 23 times faster than a bullet – to stay on a circular path around the Earth. To reach such speeds requires large quantities of expensive rocket fuel: typically the cost of putting an object into orbit is many thousands of dollars per kilogram. Anything that can help to bring this cost down is an attractive proposition, so the fact that objects on the equator are already moving eastwards at 1,670 kilometres per hour is of great interest to satellite manufacturers and space agencies: if a satellite is launched around the Earth in an eastwards direction it will receive a free 'boost' of speed from the rotation of the Earth, almost as if it's being flung into space by a giant catapult.

Even right on the equator, where the Earth's rotational speed is at its maximum, this boost is only a small fraction of the total speed required to achieve orbit. The further north or south you are, the lower the rotational speed and the smaller the boost but in such an expensive business every little helps – and the resultant savings in fuel can still be significant. Except for some specialized satellite applications, launching in an easterly direction certainly makes more sense than launching towards the west and having to fight against the Earth's rotation in order to achieve orbital speeds. This is why many artificial satellites, including the International Space Station, move across the sky from west to east – counter to the apparent direction of natural objects such as the Sun, Moon and stars. It is also one reason why spacecraft launch sites are often located as close to the equator as possible: the Americans launch from Cape Canaveral in Florida, the European Space Agency from Kourou in French Guiana, and the Russians from Baikonur in Kazakhstan, in the south of the old Soviet Union. There are even proposals for floating launch pads that could be anchored in the equatorial oceans to extract the maximum financial benefit from the Earth's spin.

The same trick can be used in reverse when attempting to bring spacecraft safely to the ground, either on Earth or on other planets and moons of the Solar System. Here, mission engineers have the opposite problem: the enormous speed required to keep the spacecraft in orbit – or to send it across the vast distances of interplanetary space – now needs to be got rid of as quickly as possible to prevent the craft from burning up in the atmosphere or hitting the ground like a missile. Rockets can be fired in reverse to decelerate the craft – but this requires additional fuel. Parachutes can be deployed to create atmospheric drag – but parachutes take up valuable room on a spacecraft, and if the atmosphere at the destination is thin or non-existent then their usefulness will be limited. However, if the craft is brought down in an easterly direction over the planet's equator, then the ground itself will be moving in the same direction beneath it, significantly reducing the relative amount of speed that needs to be lost. The appeal of this tactic can be seen in the locations of the various lander and rover missions to Mars, most of which are situated around the Martian equator.

The Earth's rotation can also be put to use in our telecommunications and weather monitoring networks. At an altitude of 35,786 kilometres above the equator, the speed required to stay in orbit is 3.07 kilometres per second. This means that a satellite will travel once around the equator in exactly the same time that it takes the Earth to make one complete turn on its axis. If the satellite is moving in an easterly direction – the same as the direction of the Earth's spin – then it will move precisely in step as the planet rotates beneath it. From the point of view of someone standing on the ground, the satellite will appear to hover permanently over the same equatorial location, as if it was tethered to the top of a very tall tower. At such a high altitude, it will be directly visible from far and wide, making it perfect for relaying signals around the globe and gathering data and images over large areas of the Earth's surface.

The spin even affects the shape of our planet: because it is moving more quickly than the polar regions, the equator bulges outwards so that, rather than a perfect sphere, the Earth presents a slightly flattened profile – a shape known as an 'oblate spheroid'. As we've seen, measured around the equator the circumference of the Earth is 40,075 kilometres; but, if the circumference is measured taking in the poles, it is only 40,008 kilometres – a small but significant difference of 68 kilometres.

This 'squashed' shape has some fascinating consequences, although of minor importance in the grand scheme of things. Everyone learns at school that the highest point on Earth is the summit of Mount Everest, towering 8.8 kilometres above local sea level. But this is not the furthest point from the centre of the Earth. Situated further south than Everest and very close to the equator, the volcanic peak of Chimborazo in the Andes of Ecuador is only 6.3 kilometres above local sea level but, because of the Earth's bulging waistline, it is over 6,384 kilometres from the centre of the planet – more than 2 kilometres further than the summit of Everest.

It is also the case that you weigh slightly less at the equator than you do at the poles. This is due to a combination of two effects, both of which are ultimately a result of the Earth's spin. Because objects at the equator are further from the centre of the Earth, they experience a weaker gravitational field and are pulled downwards with slightly less force. And because the rotational speed is higher at the equator than at the poles, objects will also experience a larger outward centrifugal force, which works against the inward force of gravity. (Physicists will tell you that 'centrifugal force' is a misnomer, as it isn't really an outward force at all – rather, it is the result of keeping an object on a circular path when its natural inclination is to fly off in a straight line. Nevertheless, for the rotating object itself it certainly feels as though a force is pushing it away from the centre of the circle – as anyone who has been on a spinning fairground ride will know.) Overall, the difference amounts to around 1 per cent of your weight but, alas, heading to the equator won't actually make you any slimmer: it will simply reduce the force with which your body presses down on the weighing scales.

Our planet's flattened profile is even reflected in the way we measure length and distance. After the French Revolution in 1789 the republican government decided to adopt a new, more rational system for measuring length – it was to be decimal (making it simpler to work with) and based reassuringly on a fundamental, unchanging property of the Earth itself. The result was the metre (French for 'measure'), which was originally defined as a ten-millionth of the distance between the north pole and the equator – in other words, a ten-millionth of a quarter of the circumference of the Earth as measured through the poles. This would give a polar circumference of 40,000,000 metres (or 40,000 kilometres), although the flattening and small-scale lumpiness of the Earth is quite hard to model accurately and so these eighteenth-century calculations differ slightly from the modern value. Nevertheless, the length of a metre would be significantly different if it had been based on a ten-millionth of a quarter of the equatorial circumference rather than the polar one. Metres, and all their subdivisions and multiples of kilometres, centimetres and millimetres, would be slightly longer than the measures we're used to. The actual distances between two points on the map would still be the same but we would describe them differently: London and New York are 5,577 kilometres apart by our current system, but with an alternative, longer 'equatorial metre' as the basis for distance this would be 5,568 kilometres.

The Earth's spin has consequences for our biology as well as our technology and systems of measurement. As diurnal animals, we humans are closely tied to the Earth's daily cycles of light and darkness. Artificial lighting may have allowed us to bend the biological rules a little but for all our talk of 24-hour lifestyles we still tend to be awake during the hours of daylight and to rest when it's dark – just as our ancestors have done for millions of years. The spin has even more profound consequences for the climate and, ultimately, the habitability of the planet itself.

At any particular moment, half of our planet's surface is turned towards the Sun and is receiving its heat and light, while the other half faces out into space and is radiating its warmth back into the freezing darkness – which is why the night is generally cooler than the day. But the Earth is continually turning so no part of the surface is being either permanently heated or cooled. Our planet's atmosphere plays a huge role in smoothing out the temperature differences between day and night, particularly as it contains powerful greenhouse gases that help to retain the Sun's warmth. Energy acquired by the ground during the day is released in the form of infrared radiation but much of this is re-absorbed by gases such as carbon dioxide, methane and water vapour in the atmosphere, delaying its journey back out into space. Meanwhile, the water of the oceans also acts as a giant heat storage system, ensuring that warmth acquired during the day is released slowly throughout the hours of darkness.

We can see just how much the Earth's thick blanket of air and water help to even out the temperature differences between daylight and darkness by looking to our nearest neighbour in space, the Moon. Our natural satellite is the same distance from the Sun as the Earth so on average its surface receives around the same amount of solar heat per square metre. However, the Moon spins on its axis much more slowly than the Earth and a 'lunar day' lasts around 27.5 Earth days (we see this long transition between lunar daytime and lunar night as the changing phases of the Moon). Without an atmosphere to retain the Sun's heat, temperatures during the two weeks of daylight rise to over 120 degrees Celsius, while during the two-week period of darkness they rapidly drop to a chilly -150 degrees. There are even some deep craters in the Moon's polar regions where sunlight never penetrates, and with no air to transfer heat the temperature here remains permanently at -150 degrees Celsius, even when the surrounding surface is bathed in sunlight. The constant low temperature means that such craters are prime sites to look for frozen water trapped in the lunar soil – which could be a useful resource in future attempts to colonize the Moon.

Astronauts on the International Space Station (ISS) also have to contend with huge fluctuations in temperature as they race around the Earth every 92 minutes. When the ISS is in the Earth's shadow its outside temperature drops to around -150 degrees Celsius. When it emerges again into sunlight 46 minutes later, the situation becomes even more extreme: the side of the station facing the Sun is rapidly heated to over 120 degrees Celsius while the side facing away from the Sun remains at around -150 degrees. Insulation is key to keeping the interior of the ISS buffered against these rapid external changes and the station is wrapped in layers of reflective Mylar, similar to the metallic blankets distributed to marathon runners after a race. In fact, so successful is this insulation that the biggest problem is getting rid of the excess heat generated inside the ISS by the station's electronics and the bodies of the six crew members.

Other planets undergo even harsher daily extremes of temperature. Orbiting the Sun at just one-third of the distance of the Earth, Mercury has only a vanishingly thin atmosphere to redistribute the huge amounts of heat that it receives. Here, the surface temperature ranges from over 500 degrees Celsius during the day to -173 degrees during the night although, as on the Moon, there are believed to be 'craters of perpetual darkness', which have some of the lowest temperatures in the Solar System and – perhaps – also contain deposits of frozen water (*see* Water Everywhere, page 52). Venus is only slightly further from the Sun than Mercury and rotates on its axis only once every 243 Earth days, but the planet's thick atmosphere – 93 times the mass of Earth's – is composed mostly of carbon dioxide and the heat-storing properties of this greenhouse gas ensure that there is barely any difference in temperature between the day and night sides of the planet – or indeed between the equator and the poles.

Despite Earth's atmosphere and oceans acting like an insulating blanket, helping to store the heat received during the day, even on the lit day-side of the Earth, the Sun's heat is not evenly distributed. Far more heat is received at the equator than at the poles and this difference is the driving force behind our planet's weather, as air masses and ocean currents move around in an endless attempt to redistribute the heat evenly across the globe. The Earth's rotation plays a key role in this process too, shaping the movement of air and water through a powerful phenomenon known as the Coriolis effect. Weather and climate are ultimately powered by energy from the Sun but the forms they take are, to a large extent, a product of our spinning planet.

In the tropics, the land and sea are heated by the full force of the Sun. In turn, they warm the air above them, which then begins to rise. At an altitude of around 15 kilometres, this warm air begins to flow north and south, away from the equator, while at ground level more air is drawn in towards the equator to replace it. These are large-scale movements over thousands of kilometres and, as the air masses move north and south, the difference in the rotational speed between the equator and higher latitudes starts to have an effect. Air moving north or south away from the equator has a greater eastward speed than the latitudes to which it is moving, and it therefore drifts towards the east relative to the ground as it goes. Meanwhile, air flowing towards the equator has a lesser eastward speed than its destination and therefore drifts towards the west. This sideways deflection of the northern and southern flows of air is a manifestation of the Coriolis effect. It exerts a profound effect on the Earth's atmosphere and oceans and, through them, has helped to shape human history and culture.

By the time it reaches latitudes of around 30 degrees north and south, the high-altitude air has been deflected so much towards the east that it is now flowing parallel to the equator rather than away from it – and it begins to sink back to the ground. As it rose over the equator, its cargo of moisture was released as rain, sustaining the lush band of tropical rainforest that encircles our planet. Now the sinking air is very dry, so at latitudes of around 30 degrees north and south of the equator great deserts such as the Sahara also form rings around the Earth.

On reaching ground level, the air is drawn back towards the equator, and again the Coriolis effect comes into play – this time deflecting it towards the west. Looking down on the Earth from space, the Coriolis effect causes air masses in the northern hemisphere to curve in a clockwise direction while in the south the deflection is anticlockwise. In the northern hemisphere, this circulation pattern is responsible for the steady winds that blow across the Atlantic and Indian Oceans in a southwesterly direction for much of the year. These are the Trade Winds, prized over the centuries by mariners for their reliability. They have played an important role in many episodes of human history – including the European colonization of the Americas and the rise of the British Empire in India – and they are a direct result of the Earth's spin.

The effect is not just limited to the tropics. At higher latitudes, the same process of rising, falling and deflection of air via the Coriolis effect produces two bands of atmospheric circulation that define the northern and southern temperate zones. Two further systems form caps over the planet's polar regions. Without the Earth's spin to shepherd the moving air masses on their curving paths, the familiar climatic division into tropical, temperate and polar zones would be somewhat different.

As well as the underlying structure of the Earth's climate system, the Coriolis effect also plays a role in the short-term variations that make up our weather. It determines the direction of airflow around the low-pressure and high-pressure systems that bring either clouds and rain or spells of fine weather. The planet's jet streams – high-altitude 'rivers' of air that spiral around the globe and influence everything from airline routes to the path of storm systems – are also shaped by the Coriolis effect. Even the famous British obsession with the weather might be thought of as an indirect result of our spinning planet: the British Isles sit right on the boundary of the northern temperate and polar zones, resulting in their endlessly changeable climate.

The movement of air over water is one of the main drivers of currents in the oceans, and so the Coriolis effect also makes its mark on the large-scale circulation of water around the planet. These circulation patterns, known as 'gyres', can be thousands of kilometres across and, like the large-scale motions of the air, they rotate in a clockwise direction in the northern hemisphere and anticlockwise in the south. Because water is very good at storing heat, the gyres play a major role in transporting warmth from the equator to higher latitudes. And because warm water is the energy source for tropical cyclonic storms, such as hurricanes and typhoons, the Coriolis effect is also the underlying reason why these most violent of weather phenomena spin in opposite directions in the northern and southern hemispheres.

As well as weather and climate, the ocean gyres have also left their mark on human history and culture. Alongside the trade winds they have helped to shape global shipping routes and through them human commerce, exploration and migration. These vast circulations of water also play a crucial role in fertilizing the oceans with nutrients, driving the food chains on which our fishing industries depend. Their contribution is significant: the cool Humboldt Current, which flows up the west coast of South America, covers just 1 per cent of the ocean's surface but accounts for around 20 per cent of global fish catches. Even seafood owes a debt to our planet's spin.

The influence of the Coriolis effect is not confined to our atmosphere and oceans: it makes its presence felt whenever anything moves a significant distance between the equator and the poles. For example, during the First World War, German forces employed giant 34-metre-long cannons, known as 'Paris Guns', to bombard the French capital from a distance of 120 kilometres. Calculations of the shells' trajectories had to include the Coriolis effect to ensure that they hit their target rather than curving off to the side. The effect is even more pronounced over the intercontinental distances traversed by modern ballistic missiles.

Air and water are not the only fluids on Earth that undergo large-scale motions. Deep beneath our feet is the ocean of molten iron and nickel that makes up the planet's outer core. As turbulent eddies of liquid metal rise from lower, hotter regions of the core, the Coriolis effect arranges them into great rolling flows that spiral parallel to the planet's rotation axis. As metal is a conductor of electricity, these flows act as a dynamo, generating a vast magnetic field that extends for tens of thousands of kilometres beyond the Earth itself and surrounds the planet with a shell of magnetism.

It is this magnetic field that orients the needle of a compass, pointing it in the direction of the pole, but its importance goes far beyond its convenience for navigators. For billions of years, it has shielded our planet's surface and atmosphere from the twin dangers of the solar wind – subatomic debris blasted out by the Sun – and cosmic rays – high-speed particles from the distant reaches of our galaxy and beyond. As forms of radiation, these particles would pose a direct threat to living cells if they ever reached the surface in sufficient quantities. The Earth's atmosphere, including the protective ozone layer, would also suffer, being gradually stripped away by the onslaught of the solar wind (*see* Sunshine, page 113). A similar process is thought to have contributed to the erosion of the Martian atmosphere over billions of years, converting the Red Planet from a warm, wet world to the frozen desert that we see today. Via the Coriolis effect and the molten dynamo of the outer core, our planet has spun itself a protective net of magnetic field lines. Without it, Earth's complex biosphere would be fully exposed to the harsh radiation environment of space, and

so this protective barrier is perhaps the most important consequence of our planet's spin.

However, despite its profound influence over so many aspects of
life here on Earth, there is one phenomenon for which the mighty
Coriolis effect can't claim responsibility: the direction in which
water swirls down the plughole when you empty your bath. It is
a myth that draining water always swirls clockwise in the northern hemisphere and anticlockwise in the south. It doesn't. Across
the width of a bathtub any difference in rotational speed due to
the Earth's spin is tiny, and instead the direction of flow will be
determined each time you empty your bath by random effects
such as currents in the water and the shape of the bathtub itself.

The Earth's spin seems like a banal and everyday fact, and it is
easy to take for granted the fact that we live on the surface of a
revolving globe. But it is worth remembering that, as recently as
the seventeenth century, the idea that the Earth could be spinning was a profoundly shocking one for many people. Perhaps
we should try to recapture a little of that amazement, because
without its spin our planet would be a very different place
indeed. A day would last for an entire year, with one half of the
planet exposed to the heat of the Sun for months on end, while
the other was plunged into icy darkness. The Earth would be
more perfectly spherical, lacking its equatorial bulge, and, with
no spin to send air currents and oceanic gyres coiling across the
globe, our climate and weather would be radically different.
Worst of all, without the swirling of liquid iron in the Earth's
outer core, our planet would lack a protective magnetic field: the
atmosphere would be inexorably stripped away by the harsh

solar wind and the surface would be bombarded with dangerous levels of radiation. A spinning planet is a habitable planet. We may not be able to feel it, but without the Earth's constant turning, we probably wouldn't be here at all.

Solid Ground

Even though we know that the Earth is hurtling through space, spinning on its axis and wobbling as it goes, it still feels stable and motionless beneath our feet. As we have seen, this is because we are actually moving along with it and thus we don't feel our own motion, but so strong is our impression of the Earth's immovability that the idea of 'solid ground' remains a powerful metaphor for reliability and permanence.

In fact, this sense of solidity is even more misleading than it seems – as anyone who has experienced an earthquake will know. The ground beneath our feet is actually just a thin crust a few kilometres deep, floating on top of a layer of hot, viscous rock known as the mantle, which extends down to a depth of almost 3,000 kilometres. Although technically a solid, the rock of the mantle exists at such high temperatures and pressures that it is able to deform and flow – more like putty than the rigid crystalline minerals that we are used to at ground level. The vast currents and upwellings of mantle rock push and pull at the overlying crust, driving the slow movements of the continental plates. When the resulting stresses and strains overwhelm the strength of the rigid crust, any excess pressure is released in the sudden violence of a tremor or quake. This shaking motion of the Earth's surface can cause death and destruction on a terrifying scale but even a relatively minor tremor can still be a profoundly disconcerting experience, as the ground shakes and sways with unpredictable motion.

How do we know what lies thousands of kilometres beneath our feet? Our deepest mineshafts only penetrate a few kilometres into the crust, barely scratching the planet's surface – so digging isn't the answer. At active volcanic sites geologists can analyse molten rock welling up from the mantle, and with seismic detectors geophysicists can study the way that the shockwaves generated by earthquakes are transmitted through the interior of the Earth and back to the surface – both revealing details of the density and composition of our planet's interior. But, surprisingly, one of the best methods that we have for exploring the depths of the Earth actually comes not from looking down, but from looking up.

Very Long Baseline Interferometry, or VLBI, is a technique developed by radio astronomers to study distant celestial objects in extremely fine detail. A general rule in astronomy is the larger the telescope the finer the details it can pick out in the sky – and the more information astronomers can glean about the objects they are studying. Bigger is therefore often better when it comes to telescope design, but all too often the grandiose ambitions of astronomers run up against the hard reality of engineering limitations and – inevitably – funding constraints. Radio astronomers have found a way of getting around this, not by spending more money on new telescopes but by cleverly linking together the ones they already have. By pointing radio dishes all around the world at the same patch of sky, then carefully combining their signals, astronomers can mimic the performance of a single giant radio telescope with a receiving dish up to several thousand kilometres across.

VLBI can reveal detail up to a thousand times finer than that achieved by instruments like the Hubble Space Telescope and it has contributed enormously to our understanding of cosmic phenomena such as quasars, black holes and the birth and death of stars. But, in order for it to work, the technique has one crucial requirement: the positions of the individual telescopes must be known with extreme precision.

It is this positional precision that makes VLBI such a useful tool for geophysicists studying the interior of the Earth. Effectively the VLBI process can be turned on its head: just as navigators and cartographers in previous centuries took fixes on the stars to calculate their positions down here on Earth, so the radio dishes of a VLBI network can take fixes from distant radio sources to pin down the precise positions of the dishes around the globe. The method uses quasars – extremely distant point-like radio sources whose celestial positions have already been well determined by decades of observations. By comparing measurements taken over a period of time it allows geophysicists to track tiny changes in the locations of the individual radio telescopes here on Earth – even down to scales as fine as a few millimetres. Using their telescopes in this way, scientists can directly measure the motions of the great tectonic plates that make up the Earth's crust, and detect the uplift or subsidence caused by the motions of the mantle rocks deep below. These motions give hugely valuable insights into the workings of the planet's turbulent interior, and so a technology developed to explore the depths of outer space has ended up helping us to investigate the inner depths of our own world.

Beneath the slowly flowing rocks of the mantle, our planet becomes even less like the solid lump of rock that we imagine it to be. The outer part of the Earth's core is an ocean of liquid iron and nickel about 2,000 kilometres deep, with a temperature of over 4,000 degrees Celsius. It is here that the Earth's magnetic field has its origins: the planet's spin causes the eddying motions of the liquid metal to act like a giant dynamo. In man-made dynamos, such as the ones attached to bicycle wheels to power a headlight, a rotating magnet is used to generate an electrical current in a conducting wire, but in the outer core of the Earth it is thought that a similar process acts in reverse, with the movements of the conducting metal creating a current that in turn generates a web of magnetic field lines that extend up through the surface of the planet and tens of thousands of kilometres out into space (*see* Spinning Around, page 265). For hundreds of years, humans have made active use of the planet's magnetic field as an invaluable navigational aid, taking advantage of the way it aligns a magnetic compass needle towards the poles. We – and all other living things on Earth – have also derived a more fundamental benefit from it for billions of years: by shielding our planet from the worst of the radiation blasted out by the Sun and other objects in space, the magnetic field plays a crucial role in keeping conditions on the surface suitable for life (*see* Sunshine, page 113).

Only at the very centre of our planet can we find anything approaching true solidity, but even here the definition of solid is unlike anything we're used to on the surface. The Earth's inner core is a giant ball of iron and nickel around 1,220 kilometres across – about 70 per cent of the size of the Moon. With a temperature of 5,400 degrees Celsius – as hot as the surface of the Sun – the metal of the inner core is even hotter than the liquid shell that surrounds it, but the colossal weight of the over-lying layers of metal and rock exerts a crushing pressure, more than 3 million times greater than air pressure at sea level, which is enough to keep the iron and nickel alloy 'frozen'. If the surrounding layers of the planet were somehow removed, this pressure would be released and the inner core would explode into incandescent vapour.

It is extremely challenging to try to reproduce the intense temperatures and pressures that exist at the centre of the Earth in a laboratory, so we still don't completely understand how substances behave under such formidable conditions. However, we can say for sure that the lump of iron-nickel alloy at our planet's core must be rather different from the familiar metals that exist in the cool, low-pressure environment here on the surface. Studies of the way that seismic waves from earthquakes pass through the inner core seem to show that it has a compli-cated structure, containing giant iron crystals aligned in a north-south direction. There are even hints that the inner core may itself be divided into outer and inner regions, both with slightly different properties. Whatever its true structure, it is slowly growing in diameter, at a rate of around 1 millimetre every year, as more metal crystallizes out from the liquid layers above: as heat is slowly conveyed outwards through the rocks of the mantle this cools the planet's interior, allowing the liquid metal of the outer core to crystallize out onto the solid inner core.

The solid ground in which we place our faith is really just a brit-tle skin of minerals resting on the putty-like rock of the mantle, which in turn floats on the metallic ocean of the outer core. At the very heart of our world is a ball of superheated metal crystals, which itself remains solid only by virtue of the unimaginable pressure of the metal and rock above. The turbulence of the fluid interior drives the motions of the tectonic plates and shapes the geology of the crust, dictating everything from the outlines of the continents to the distribution of minerals on which our technology and agriculture depend. The movement of the continental plates affects us in an even more fundamental way: as two plates jostle together, their edges sliding over and under one another, water and minerals from the surface are 'subducted' by the sinking plate and dragged down into the upper mantle, to eventually emerge again at the surface via volcanic vents. This constant cycling of material between the atmosphere, oceans and rocks is one of the great geological support systems that help to keep our planet habitable.

The internal motions of rock and metal are driven by the upward convection of heat, starting from the inferno of the inner core and slowly working its way, via swirling currents and turbulent eddies, to the surface thousands of kilometres above. But why is the interior of the Earth so hot in the first place? Geologists think that there are two main sources for our planet's inner heat – and ultimately both of them have extraterrestrial origins.

The first heat source is a direct relic of the Earth's violent birth 4.5 billion years ago. As rocky fragments and large asteroids crashed into the growing planet – often at speeds of several kilometres per second – the successive impacts heated its rocks to melting point. In the chaos of the early Solar System, so much debris was raining down onto the embryonic world that heat was being delivered faster than it could be radiated away into space. As the planet grew, its increasing gravity also squeezed and compressed the interior, heating and melting it still further. Heavy elements such as iron and nickel sank into the depths of the molten world while lighter silicate materials – the building blocks of many types of rock – floated to the surface. The early Earth became a cauldron of liquid rock and metal, with a fluid core and a surface ocean of glowing magma.

SKY METAL

The Earth's metallic core is almost certainly forever beyond our reach. Buried beneath thousands of kilometres of superheated rock, the temperatures and pressures at such depths would be enough to melt and crush any digging equipment we could devise. Even simulating such extreme conditions in laboratories here on the surface is an extremely challenging prospect. But, surprisingly, we do already have samples of crystallized metal that formed under conditions similar to those at the centre of the Earth. They come not from deep beneath our feet but far out in space.

Meteorites are fragments of space debris – mostly originating in the asteroid belt between the orbits of Mars and Jupiter – which survive their incandescent passage through the Earth's atmosphere to land more or less safely on the ground. The majority of meteorites are composed of rocky minerals but around 6 per cent of them are made of metal – a crystalline alloy of iron and nickel, often with trace amounts of other elements such as cobalt.

How did chunks of metal come to be floating around in space? The answer is that they are fragments of the cores of long-lost 'protoplanets' – embryonic worlds that began to form early in the history of the Solar System but were smashed apart in violent collisions with other objects before they could attain full planet status. Some of these doomed objects had already grown to a size at which their interiors became molten, allowing the process of differentiation to occur in which the heavy elements like iron and nickel sank to the centre while lighter, rocky material floated to the top.

With typical sizes of a few hundred kilometres across, most of these objects were much smaller than the Earth and so they were also able to cool much more rapidly, allowing their cores to lose their initial heat and crystallize into solid balls of metal. This cooling process would still have taken millions of years, but when it was complete the early Solar System – and particularly the asteroid belt – remained a crowded place. Chemical analysis of metal meteorites that have fallen to Earth indicate that they derive from at least 50 distinct parent bodies, which implies that there must have been many large, differentiated objects in the asteroid belt at this time. High-speed encounters between these objects were common and, one by one, the remaining protoplanets collided with each other, shattering into clouds of rock and metal fragments.

Apart from the fully fledged inner planets Mercury, Venus, Earth and Mars, there are few survivors from this ruthless process of elimination, and the asteroid belt is littered with the rubble of the unlucky majority. Three prominent exceptions are the largest asteroids, Ceres, Vesta and Pallas, which have diameters between 500 and 1000 kilometres and together account for more than half of the total amount of material in the asteroid belt. NASA's Dawn spacecraft paid a visit to Vesta in 2011 before heading on to Ceres in 2015 – in both cases revealing roughly spherical worlds, battered and scarred by billions of years of collisions with smaller objects. Vesta, in particular, seems to have survived only by the skin of its teeth: the southern hemisphere has been gouged out by two enormous impact craters, giving it a distinctly lop-sided appearance. These impacts would have littered the Solar System with fragments of Vesta's rocky crust and mantle, and indeed more than a thousand stony meteorites have been found on Earth with chemical signatures that match those of the asteroid, giving geologists invaluable samples of a world frozen in the early stages of becoming a full-grown planet.

Vesta, Ceres and Pallas are the exceptions, scarred but relatively intact. The majority of the protoplanets were smashed to pieces, cores and all, releasing their differentiated contents back into space. This is how pieces of metallic iron and nickel came to be floating around the Solar System today – and when they fall to Earth, these metal relics provide us with an insight into the secret heart of our own world.

As you might expect, iron that solidified in the heart of a protoplanet is somewhat different from metal forged here on Earth. One characteristic of meteoritic iron becomes apparent when the metal is sliced and lightly etched with acid. Intricate patterns of interlocking metal crystals, formed from different alloys of iron and nickel, appear on the cut surface. Known as Widmanstätten patterns, these beautiful criss-crossing structures give iron meteorites an aesthetic as well as scientific appeal: they are named after Count Alois von Beckh Widmanstätten, the director of the Imperial Porcelain Works in Vienna, whose fascination with the properties of different materials led him to experiment with a meteorite sample in 1808.

Structures like these never occur in the manufactured iron that emerges from our foundries: the separation and crystallization of the different iron-nickel alloys requires a steady cooling process lasting millions of years. They are therefore sure signs of an extraterrestrial origin and a testament to the long period this metal spent buried in the heart of a protoplanet while the Earth itself was still forming.

Widmanstätten's name is also attached to an asteroid and a lunar crater so his astronomical status is secure. However, as often happens in science, the credit for discovering the crystalline structure of meteoritic iron does not belong to him alone. The geologist G. Thomson discovered it entirely independently in 1804 but, working in Naples during a period of political instability, he had difficulty bringing his discovery to the attention of the international scientific community. Civil war, the murder of a courier carrying his manuscripts and Napoleon's invasion of Italy all intervened, and Thomson died in 1806 without achieving recognition for his finding – a reminder that scientists rarely work in isolation from the misfortunes and inconveniences of history.

Iron meteorites also contain another reminder of their ancient origins. Just as motions in the Earth's liquid outer core are thought to generate our planet's magnetic field (*see* Spinning Around, page 265), so a similar dynamo effect seems to have been in operation in the centres of proto-planets while they too were molten. When the protoplanet cores had completely cooled and solidified, their magnetic fields died away but evidence of their presence remained frozen in place in the form of tiny magnetized crystals of nickel-iron alloy. Like the data encrypted on the hard drive of a computer, these crystals preserve a record of the core's magnetic properties, from its initial liquid state to its final solidification, giving us an intriguing glimpse of the past, present and ultimate future of our own planet's core.

As well as giving us invaluable insights into the inaccessible centre of the Earth, iron meteorites have also exerted a cultural influence on human societies for thousands of years. As everyone knows, metallic iron exposed to the air quickly begins to rust, reacting with oxygen to form orange iron oxide. So strong is iron's affinity for oxygen that virtually all the iron present in the Earth's crust is locked away in oxide ores, while deposits of pure metallic iron are almost non-existent. It was only at the beginning of the Iron Age around 3,200 years ago that humans developed the high-temperature smelting technology necessary to extract metallic iron from naturally occurring ores. The ready availability of this versatile new metal helped to drive developments in agriculture, social organization and even art, and is regarded by archaeologists as a significant turning point in human history.

And yet a handful of metallic iron artefacts have been discovered dating back to the Bronze and Stone Ages, hundreds or even thousands of years before the technological breakthroughs that heralded the age of iron. Without the technology to smelt iron ore, there is only one possible source for the metal used by these ancient people: meteoritic iron from space. Certainly anyone witnessing the fiery descent of an iron meteorite would be in no doubt as to the unusual nature of this shiny stone from the heavens. But for Stone Age people attuned to every aspect of their environment, and the usefulness of the materials it contained, even a piece of meteoritic iron that had fallen many years earlier would have been recognized as a highly significant find.

The use of meteoritic iron occurred in many ancient cultures across the world and the extreme rarity of this material meant that it was often made into high-status artefacts or even venerated in the form of religious objects.

In 1911, beads worked from meteoritic iron were discovered by the great Egyptologist William Finders Petrie in an Egyptian grave dating to 3,200 BCE. They were part of a necklace that also contained beads of gold, lapis lazuli and other precious materials, indicating that this unique and remarkable metal was highly prized. The discovery of an ornate dagger with a blade of meteoritic metal in the tomb of Tutankhamen shows how iron continued to be venerated by the Ancient Egyptians throughout the Bronze Age.

In Europe, stories of magical swords forged from meteorites are found in folktales and have even made their way into the invented mythology of J.R.R. Tolkein's Middle Earth. In more recent times, iron meteorites have been put to good use by traditional societies whose hunter–gatherer lifestyles lacked the facilities for smelting iron ore. For centuries, the Nama people of southwest Africa have been making tools from scattered fragments of the Gibeon meteorite, while the Inuit of Greenland made arduous overland trips to the impact site of the Cape York meteorite to collect iron from which to make harpoons. What the Inuit thought when their invaluable source of iron was carted off to a museum in the 1890s is another matter entirely.

The wide availability of iron smelted from terrestrial ores has not entirely removed our desire to fashion meteoritic iron into objects for our own use. Convenience rather than necessity was perhaps the reason why iron from the Cranbourne meteorite was made into a horseshoe and a hotplate in nineteenth-century Australia, but in the twenty-first century scientists are intrigued by the unusual magnetic properties of meteoritic iron crystals. They may hold the key to the manufacture of improved magnets for green technologies such as electric cars and wind turbines. There is even a thriving trade in jewellery and ornaments crafted from this heavenly metal. Astronomers and geologists may wince to see such precious specimens worn as decoration, but perhaps it goes to show that we value these visitors from space in more ways than just the scientific.

Eventually, after a few million years, the young Solar System began to calm down. Most of the original material from which it had formed – gas, dust, ice and rocky fragments – had either been incorporated into the Sun and planets or driven away by the increasing radiation of the newborn Sun, or else – like the rocky debris of the asteroid belt or the icy fragments of the distant Kuiper Belt and Oort Cloud – corralled into well-defined orbits, from which the gravity of the planets would have difficulty dislodging them. As the impact rates on the Earth and the other rocky planets declined, these worlds were at last able to radiate heat away faster than they received it – and slowly they began to cool. Gradually, the Earth's surface solidified into a rigid crust, while deep in the metallic core the long process of crystallization began.

The rate at which the Earth and the other rocky planets were able to lose their primordial heat depended to a great extent on their size. Heat is lost through an object's surface and because small objects have more surface area relative to their volume (area depends on the square of an object's size, while volume depends on its size cubed), they lose heat more effectively than larger objects – and can therefore cool more quickly. (This is one reason why animals that live in cold climates tend to be bulkier than their tropical counterparts, because it helps to minimize heat loss through their skin.) By the present day, 4.5 billion years after the Solar System first formed, many of its smallest members – the asteroids – have been able to cool down enough for their interiors to become entirely solid. Larger objects such as the Moon and Mercury are thought to be mostly solid with perhaps a liquid component to their cores, while Mars may still possess enough fluidity in its mantle rocks to be volcanically active at a low level. Meanwhile, the Earth and Venus, the largest of the rocky planets, still retain significant amounts of their primordial heat.

However, the Earth's relatively slow cooling rate is not enough on its own to explain why the interior of our planet is still so hot: far hotter than it should be if it had just been quietly losing its initial store of primordial heat for 4.5 billion years. Somewhere deep inside there must be an additional heat source.

When our planet formed out of the nebula that gave birth to the Solar System, it incorporated a cross-section of the different atoms and molecules that made up this primordial cloud of gas and dust particles. Hydrogen and helium were by far the most common elements but in the warm conditions of the inner Solar System, close to the newborn Sun, they were too light and fast moving to be easily held by the comparatively weak gravity of the Earth. Our planet, along with the other inner planets Mercury, Venus and Mars, is therefore largely made up of the remaining heavier atoms and molecules: metals like iron and nickel, the silicates and other rocky compounds that make up our minerals, and the nitrogen, oxygen and carbon dioxide of our atmosphere – along with the huge variety of other substances that we find on Earth today.

Mixed in with these were small amounts of radioactive elements: atoms whose central nuclei are unstable and prone to decay. When they do, the nuclei break apart to produce 'daughter' elements along with a burst of energy, which can be manifested as heat. This 'radiogenic heat' is thought to be the second contributor to the Earth's internal heat budget, contributing perhaps half of the total – though the exact proportion is still a matter of active research and debate among geophysicists.

Scientists think that four particular radioactive elements – uranium-238, uranium-235, thorium-232 and potassium-40 – are involved in radiogenic heat production inside the Earth today, although others, now completely decayed away, may also have contributed in the distant past. They are concentrated in the rocky mantle rather than the core. Each unstable atomic nucleus is like a tiny package of stored energy, waiting for the randomly determined moment of its decay to release its cargo as heat.

This splitting apart of atomic nuclei with an accompanying release of energy is known as nuclear fission and we humans have learned to use it ourselves in a number of ways. As spacecraft venture to the outer reaches of the Solar System, the available supply of solar energy dwindles as they get further from the Sun. To overcome this, they are often equipped with 'nuclear batteries' that use the decay of a radioactive element to produce heat, which can then be converted into electricity to power the craft. Back on Earth, the fission of radioactive nuclei is harnessed to generate significantly greater amounts of power in our nuclear reactors. Here, uranium-235 is artificially purified and concentrated in sufficient quantities so that the decay of each atomic nucleus is able to trigger the further decay of at least one of its neighbours, producing a self-sustaining chain reaction. The heat released by these disintegrating uranium-235 nuclei is used to create steam, which then drives turbines to generate electricity. A less controlled chain reaction, in which the fission of one atomic nucleus triggers the fission of its neighbours in a runaway cascade, is the process at the heart of our nuclear weapons. It might be disconcerting to think that about half of the Earth's internal heat ultimately relies on the same nuclear physics as an atom bomb but when we tap into subterranean heat sources to generate 'clean' geothermal electricity we are really using – in part at least – a natural form of nuclear power.

The amount of radioactive material present inside the Earth is gradually diminishing as, one by one, the radioactive atoms that were present when the planet first formed decay away. This means that Earth must have had a higher abundance of radioactive elements in the past than it does today. The decay of each individual atom – and the release of the energy it contains – is a random process, but each type of radioactive element has its own characteristic decay probability, which in turn governs how likely its atomic nuclei are to fall apart. This means that we can predict very accurately how much of a particular radioactive element will be left after a given time period – and, by putting this prediction into reverse, we can use the amount of radioactive material that we see around us now to work how much of it there must have been in the past.

From the current abundance of radioactive potassium-40 atoms and the known rate at which their nuclei decay we can calculate that the newly formed Earth must have contained around 12.5 times as much potassium-40 as it does today. Similarly, the abundances of uranium-238, uranium-235 and thorium-232 would also have been higher and geophysicists estimate that 4.5 billion years ago the total amount of radiogenic heat being released by all these radioactive elements was around four times greater than the present value.

The increased abundance of uranium-235 atoms in the past also helps to explain a strange geological phenomenon known as a 'natural nuclear fission reactor', the remains of which have been discovered in Gabon in West Africa. It seems that around 1.7 billion years ago a naturally occurring seam of uranium achieved the conditions required to produce a self-sustaining chain reaction of radioactive decay similar to a modern nuclear reactor, generating huge amounts of heat deep beneath the ground. This was only possible because the abundance of uranium-235 was around four and a half times greater than it is now. Such a nuclear chain reaction could never occur on its own today and our own nuclear reactors have to be fuelled with artificially refined and concentrated uranium-235.

But if radioactive elements are unstable why do they exist in the first place? Clearly they must have been created at some point in the past – and recently enough that there are still sufficient amounts remaining to warm the Earth today. This is indeed the case: like all atoms heavier than hydrogen and helium, the Earth's complement of radioactive elements was made inside stars that lived and died before our Solar System formed. But in order to create elements like these not just any star will do: the instability that makes radioactive atomic nuclei prone to spontaneously fall apart also makes them rather difficult to form in the first place. There is only one place where such reluctant atoms can be forged and that is in the intense violence of a supernova explosion – the death of a star more than eight times as massive as our Sun (*see* Made of Starstuff, page 6; Exploding Stars, page 319).

When such a star runs out of fuel, the delicate balance between gravity – which tries to squeeze the star together – and the outward pressure of the heat and light generated in its core is fatally disrupted. For millions of years, the star has been supporting itself by fusing together atomic nuclei to produce heavier and heavier elements, in the process releasing the energy required to counteract the inward force of its own gravity. But the end of the road is in sight as soon as the nuclear fusion reactions in the heart of the star begin to produce iron. Iron is useless as a further energy source for the star: it is the most stable element in the periodic table and fusing iron nuclei together absorbs rather than liberates energy. Deep inside the stellar core, a vast ball of iron 'ash' begins to grow, generating no new energy itself but still contributing to the star's overall gravitational pull. But this is not a stable situation, and as the iron core continues to grow, the star approaches a critical limit when gravity finally wins the battle and the core of the star implodes catastrophically. Within seconds, the density of the particles in the core of the new neutron star – now only a few kilometres across – causes the implosion to become an explosion, ripping through the star's outer layers and eventually propelling the debris out into space.

It is during these first few seconds of a supernova explosion that elements heavier than iron can be created. As we've seen, fusing together iron nuclei to form heavier elements requires energy, but energy is in extremely abundant supply in the heart of a supernova explosion. It's here that the remaining elements of the periodic table – including familiar substances like copper, zinc, tin, gold, silver, platinum and mercury – are generated in a final blaze of nuclear fusion. The creation of each new element captures a tiny fraction of the energy of the explosion and locks it away inside an atomic nucleus. Once formed, most of these new elements are stable, but among the supernova's yield of heavy atomic nuclei are also radioactive elements such as uranium and thorium. Because they will eventually decay, the energy captured by these nuclei is not locked away forever and, when they finally break apart, it can be unleashed once more.

Scientists believe that most of the radioactive elements found on Earth were created in one or more supernova explosions that took place around 4.6 billion years ago, immediately before our Solar System began to form. Today, billions of years after they exploded, a remnant of those dying stars still contributes to the heating of the Earth's interior – and also to the generation of electricity in our geothermal and nuclear power stations.

Whether primordial or radiogenic, both of the Earth's internal heat sources are ultimately extraterrestrial in origin. They date back to the chaotic birth of our Solar System – and beyond, to the violent death throes of a previous generation of stars. As this heat slowly makes its way from the Earth's deep interior up to the surface, and eventually out into space, it is on the return leg of a very long journey – one that brought it here from space in the first place.

Like a fading ember, the centre of the Earth is slowly cooling down. The heat and radioactive elements that have kept its interior in constant motion for 4.5 billion years were locked into our planet as it formed, but since the planet-building process ended, no new sources have arrived to top them up. As the primordial heat seeps out through the crust and back out into space the crystallization of the iron core will eventually be complete. Without a liquid outer core to sustain it, our planet's magnetic field will falter and die, leaving the Earth exposed to the full force of the solar wind and cosmic rays. And, as the radioactive elements inevitably decay away, the great convection currents in the mantle will also cease and the plate tectonics of the crust will grind to a halt. Even the air and sea will be affected: without the motion of the continental plates, the cycles that keep the oceans, atmosphere and rocks in equilibrium will also slowly fail.

However, this is not something that should worry us unduly: there is enough internal heat to keep our planet warm and mobile on the inside for billions of years to come. Indeed, as we shall see (Written in the Stars, page 308), long before the Earth can ever completely cool, cosmic events of a far greater magnitude will inevitably overtake our planet. One thing is for sure: the ground beneath our feet will never be truly solid.

Written in the Stars

For thousands of years, human beings have studied the heavens, hoping to gain an insight into what the future might hold down here on Earth. The apparent motions of the Sun, Moon and planets were believed to control the fortune of everyone from peasants to emperors, and the annual parade of the constellations about the sky was feverishly scrutinized for signs and portents of things to come.

The assumption that events in the heavens could have a direct influence on human affairs was not entirely illogical. After all, throughout the course of the year the Sun appears to move through the constellations of the Zodiac while down here on Earth, exactly in step, the seasons cycle from warm to cold, and from rainy to dry. Each year, at the time of the winter solstice, ancient people knew that on every subsequent day the Sun would spend longer above the horizon and the darkest and coldest period of the year would soon be behind them. Such information was extremely useful: every spring, when the bright star Sirius first became visible in the eastern sky before dawn, the Egyptians knew that the annual Nile flood was on its way and it was time to plant their crops. Even today, all human societies ultimately depend on agriculture for their survival, and being able to predict the timing and quality of the harvest is crucial for our food security.

It certainly seems as though there is a powerful connection between events in the sky and those here on Earth – and in an era when life was hard and unexpected events were generally to be feared, it is perhaps not surprising that people grasped at any clue to help them to anticipate future opportunities or avoid looming disaster. Over time, it was not just agricultural planning that was linked to the heavens but areas as diverse as meteorology, medicine and alchemy, as well as every aspect of human lives from individual personalities to the fate of nations. Ancient civilizations including the Babylonians, Chinese, Greeks and Indians observed the heavens with obsessive care and their theories of how the stars could influence human lives were taken up in the Middle Ages by Islamic and Christian scholars in Arabia and Europe. Across the Atlantic, Mesoamerican civilizations such as the Maya were also expert stargazers and here too a profound connection was made between events in the heavens and those on Earth.

In fact, the science of astronomy (from the Greek for 'the law' or 'the regulation' of the stars) has many reasons to be grateful to the mystical field of astrology ('the study' or 'the account' of the stars). The two disciplines arose alongside each other and, for thousands of years, the desire for accurate astrological predictions of future events was one of the main motivations – and sources of funding – for the detailed mapping and cataloguing of stars and planets on which modern astronomy is based. Indeed, many of the leading figures of Renaissance astronomy, including Nicolaus Copernicus, Tycho Brahe, Johannes Kepler and Galileo Galilei, were able to pursue their ground-breaking astronomical research largely because they were paid to provide astrological predictions for wealthy employers – regardless of whether they themselves really believed in astrology or not.

Throughout history, there have always been people who regarded astrology's claims with suspicion. The Roman philosopher Cicero asked why twins, who are born under exactly the same celestial circumstances, nevertheless have lives and personalities that can turn out very differently from each other. He also pointed out that if the planets were indeed much further away from us than the Moon then logically their influence on us – if any – could only be very tiny. Some Medieval theologians also disputed the idea that human fate was predetermined by the stars, as it contradicted religious views about free will and God's sovereignty over the human soul.

These dissenters were right to be sceptical, even if the basis for their doubts was sometimes as esoteric as astrology itself. Although there is a connection between the apparent movement of objects in the heavens and conditions here on Earth, it isn't quite what ancient people supposed. The two are linked not because the heavenly objects directly cause or influence events on Earth, but because they are both caused by the same thing. It is the Earth's orbit around the Sun, combined with our planet's tilted axis, that causes the Sun to appear to move around the Zodiac throughout the year and it is the same orbital motion and axial tilt that causes the seasonal changes of temperature and weather that have such a profound influence on our agriculture and all other aspects of human life. Both the Sun's apparent path and the Earth's seasonal changes are the result of our planet's motion, and that is the connection between them and the reason why they appear to change in step with each other.

It is also a combination of orbital motions – theirs and ours – that causes the planets to execute their complex paths against the background stars, while it is the Moon's orbit about the Earth, along with its orientation relative to the Sun, which gives us its monthly cycle of phases. The apparent motions of the heavens are the *result* of the Earth's circumstances – but they are not their *cause*.

When humans realized that the Earth was not the centre of the cosmos but rather a planet in orbit around a star, both the logical underpinnings of astrology and the accuracy of its predictions were called into question. Isaac Newton's explanation of gravity showed that, far from exerting some occult and mysterious influence, the planets themselves were subject to the same physical forces as objects here on Earth – and these forces could be measured and calculated rather than inferred from mystical beliefs. In 1675, when construction of the Royal Observatory began in Greenwich, Astronomer Royal John Flamsteed cast a horoscope for the new building. But it is clear from Flamsteed's writings on 'The Vanity of Astrology, & the Practice of Astrologers' that he had no faith in the efficacy of this divination. Instead, it seems to have been more of a ceremonial flourish – a ritual that was expected on such an occasion and, more to the point, one which demonstrated that the young astronomer was perfectly *au fait* with the complex mathematical calculations required for the job.

In modern times, when subjected to rigorous scrutiny, claims that the movements of the planets somehow influence the course of human lives soon falls apart. Statistical studies have shown time and time again that astrology's attempts to predict the outcomes of specific events are correct no more often than one would expect from random chance alone. And, despite the overwhelming success of modern physics and biology in explaining the fundamental forces of nature and the complex interactions of the living world, there is nothing in either discipline to support the idea that the motions of distant planets could significantly affect the behaviour of living creatures.

And yet there *is* a way in which the stars can tell us what the future will bring. It stems not from the mystic associations of astrology but from the predictive power of science. We can observe how the great sagas of galactic, stellar and planetary formation and dissolution have played out over the last 13.8 billion years because the light from distant objects takes decades, centuries or even billions of years to reach us. Astronomy lets us see directly into the past and we can extrapolate from these history lessons to see how the events might play out as the future unfolds. We can also use theory as well as observation to help us map out the shape of things to come. The laws of physics govern the interactions of all energy and matter in the universe: everything from the orbits of planets to the motions of galaxies is bound by their dictates. They describe how the balancing act between the inward force of gravity and the outward pressure of heat and light keeps a star stable, how fast the star will burn through its supplies of nuclear fuel, and how the star will change and evolve once this fuel is exhausted.

Between them, physics, chemistry, geology and biology tell us how our living planet will respond to changes in its astronomical environment, and even inherently random events are not beyond the ability of science to quantify and predict. Without knowing exactly when the next large asteroid will come crashing down to Earth, from the number of asteroids and the distribution of their orbits we can still calculate the probability that such an impact will occur during the next century, the next millennium or the next billion years. Advances in computing power also enable us to construct detailed simulations, taking current circumstances and extrapolating them into the future using our understanding of physical laws. By running these simulations again and again, we can assess the range of possible scenarios and calculate the most likely outcomes.

The first inevitable astronomical changes that humans are likely to notice will take place over the next few tens of thousands of years, and they will involve something central to our lives and to our civilization: the climate. Our planet's climate depends on many factors but perhaps the most important of all is the amount and distribution of solar energy arriving over its surface through-out the year, and this in turn depends on the shape of the planet's orbit and the angle of its axial tilt. Over thousands of years the gravitational effects of the other planets cause the Earth's orbit to flex slowly back and forth from almost circular to a more elliptical shape, while at the same time the tilt of the Earth's axis – which is responsible for giving us our seasons – increases and decreases like a spinning top wobbling from side to side (*see* Inconstant Moon, page 153).

These orbital and axial changes combine in a complex way known as Milankovitch cycles, which determine how much solar energy reaches the Earth and how it is distributed over the surface throughout the year. In turn, this drives large-scale changes in the planet's climate, which periodically lead to large parts of the northern hemisphere being submerged under gigantic ice sheets. Earth is not the only world in the Solar System to experience periodic climate change driven by Milankovitch cycles: similar effects cause the polar ice caps of Mars to expand and contract over thousands of years, while the methane lakes of Saturn's moon Titan migrate towards and away from the moon's poles on a 60,000 year timescale.

Back on our own planet, the cyclical expansion and contraction of the ice sheets has been going on for the last 2.6 million years. We are fortunate that we are currently living in an interglacial period – a warmer interlude between the times of glacial maximum, when the ice sheets are at their smallest extent. But geologists define the whole of the last 2.6 million years as an 'Ice Age' because, even during warmer interglacial times such as ours, there are still permanent ice caps in the polar regions. The underlying circumstances that allow this to happen are a result of the planet's changing mix of atmospheric gases and the current positions of the continents, which shape the flow of ocean and air currents. However, against the geological and atmospheric backdrop of the Ice Age, it is the astronomical Milankovitch cycles that determine when the ice sheets will next advance or retreat.

The history of human agriculture and urban society has all taken place within the 12,000 years since the end of the last glacial period – so our civilization has never had to endure the full force of a glacial maximum. Orbital and climate models predict that Europe, Siberia and North America may continue to enjoy their current ice-free interglacial climates for perhaps another 50,000 years but – inevitably – at some point in the future the

Milankovitch cycles will do their work and the ice sheets will return, grinding down to cover much of Canada, Britain and the northern parts of Europe once again. Our current profligate burning of fossil fuels, which is pumping the atmosphere full of greenhouse gases, might actually delay the advance of the glaciers for a few thousand years by artificially keeping temperatures high. But our descendants may not thank us very much for this reprieve: the damage that global warming could do to the Earth's ecosystems is a very high price to pay merely to stave off our planet's natural climate cycles for a few millennia.

In the end, there is nothing we can do to influence the Earth's axial and orbital dance. Our planet's astronomical circumstances mean that its climate will inevitably continue to change in the future, just as it has always done in the past, and we will simply have to learn to adjust to the comings and goings of the glaciers in our Ice Age world. But, on the grand scale of things, climatic variations of this nature will come to seem a relatively minor issue: astronomy tells us that there are many other changes in store, some of them far more drastic than advancing ice sheets.

Another change that will become obvious over tens of thousands of years is the appearance of the sky itself. The Sun and every other star in the sky are all engaged in their own individual journeys around the centre of the Milky Way. The stars that make up our familiar constellations are really only temporary companions – accompanying the Sun for at most a few million years before departing on their way. The changes are imperceptible on the scale of a human lifetime but over the span of recorded history they are already noticeable. The constellations that we see today are not quite the same as they were when the ancient Greeks observed them so carefully two and a half millennia ago. In the last few decades high-precision star-mapping satellites, such as the European Space Agency's Hipparcos and Gaia missions, have measured the positions and motions of hundreds of thousands of stars, and computer software can fast-forward these motions into the future, showing us how the constellations will begin to distort as their stars – and we ourselves – continue on their individual paths through the Milky Way.

After 50,000 years, the familiar 'rectangle' of the Plough (part of the larger constellation of Ursa Major, the Great Bear) will seem distinctly flattened, while its 'handle' will bend sharply down-wards. Orion the Hunter will still retain his belt of three bright stars but his elegantly curved bow will have become an awkward zigzag. Meanwhile, other well-known constellations, including the Southern Cross and the 'W' of Cassiopeia, will be completely unrecognizable. Over millions of years, the Sun and all of the stars currently in our sky will go their separate ways as their orbits take them on different paths, and at different speeds, through the disc of the Milky Way. An ever-changing popula-tion of stars will grace the skies of the far future, and our imagi-nations will be kept busy dreaming up new names for their constantly shifting patterns.

But a changing sky is not the only consequence of the Sun's galactic journey. As our Solar System moves through the disc of the Milky Way, we will periodically pass through the galaxy's spiral arms – regions rich with dense star clusters and glowing nebulae of gas and dust where new stars are constantly being born. The spiral arms are like stellar traffic jams where stars temporarily slow down and bunch together in their orbital journeys around the Milky Way. Although the arms themselves rotate about the centre of the galaxy they do so at a different rate to the individual stars, so their stellar populations are transient, forever moving through the arms and out the other side to be replaced by other stars moving in from behind.

Every time we make one complete orbit of the Milky Way, we pass through at least three of its spiral arms – possibly more, since our information on the structure of the far side of the galaxy is as yet incomplete – and in fact we are currently moving into the galaxy's Orion Arm. It takes the Sun about 10 million years to cross from one side of the arm to the other and, despite promising some spectacular night-time vistas, these passages are fraught with danger for our planet. One hazard comes from the gravity of passing stars and giant gas clouds, which could disrupt the orbits of objects in the Sun's Oort Cloud and send showers of comets towards the inner Solar System – some of them potentially on a collision course with Earth. Meanwhile, if the Solar System's trajectory takes it directly through a particularly dense cloud of interstellar dust, it is possible that the dark dust grains could temporarily reduce the amount of sunlight reaching the Earth, with consequences for the planet's climate.

Worse still, the spiral arms are packed with short-lived, massive stars, many of which are likely to end their lives violently in supernova explosions. As we've seen, supernovae have played an important role in making our Solar System habitable, by providing many of the heavy elements necessary for the formation of rocky planets and the chemistry of life – and even perhaps by triggering the collapse of the gas cloud from which the Solar System formed (*see* Made of Starstuff, page 6). But a nearby supernova today would not be such good news: if one was to detonate while our Solar System was within about 30 light years the consequences for life on Earth could be serious.

The exploding star would be dazzlingly bright in our sky for weeks or months, casting shadows by night and visible even during the day, but the real danger would come from the first few seconds of the explosion when an intense pulse of gamma rays would be released by the core of the star as it collapsed. When they struck the Earth's upper atmosphere, the gamma rays would split molecules of nitrogen apart, converting them into nitrogen oxides and initiating a chemical chain reaction that would degrade the planet's ozone layer. The initial damage would be inflicted on the hemisphere of the Earth facing the explosion but soon the entire ozone layer would be compromised, allowing dangerous levels of ultraviolet (UV) radiation from the Sun to reach the ground. Marine ecosystems are likely to be heavily affected by the influx of UV rays; indeed, some scientists have proposed that a nearby supernova may have been responsible for a mass extinction event at the end of the Ordovician period, 443 million years ago, when around 60 per cent of marine organisms were wiped out. Even distant explosions can leave their mark: ice cores from the Antarctic show traces of nitrates at depths corresponding to the dates of historic supernovae in 1006 and 1054 (as we saw in Made of Starstuff, this latter explosion was the one that created the Crab Nebula).

The good news is that, since it formed 4.5 billion years ago, the Solar System has already made about 20 complete orbits around the galaxy, and we are still here to tell the tale. This suggests that a major disaster capable of ending all life on the planet is unlikely, but scientists still debate whether the mass extinctions that palaeontologists find in the Earth's fossil record might be linked to our regular transits through the stellar minefields of the spiral arms. If the evolutionary history of life on Earth has been periodically shaped by these close astronomical encounters, then there is no reason to suppose that the future will prove any different – and if another mass extinction was to occur, we might not be among the lucky survivors.

EXPLODING STARS

If a supernova was to explode within 30 light years of the Earth, it could be catastrophic for our planet, damaging the ozone layer and triggering a mass extinction of life. Naturally no one wants a star to explode this close to us, but astronomers would dearly love to observe a supernova from a slightly safer distance, giving them a ringside seat at one of the most spectacular light-shows in the cosmos. As we've seen (Made of Starstuff, page 6), supernovae are the source of many of the heavier elements essential for the formation of rocky planets, and even of life itself, while the shockwaves generated by their debris ploughing through the galaxy could be responsible for triggering the birth of new generations of stars. A chance to study a supernova from close range would therefore give invaluable insights into our own origins. But supernovae in our galaxy have proved remarkably elusive, and no one has witnessed one in the Milky Way for hundreds of years.

From a number of clues, such as the fraction of suitably massive stars and the amount of telltale radio and gamma radiation produced by exploding stars and their remnants, astronomers have been able to estimate the rate at which supernovae should be occurring in a galaxy like ours, and it turns out that we would expect to see somewhere between one and three every century. This number is consistent with the supernova rates observed in other, similar galaxies. For the few weeks of its existence a supernova can outshine all of the billions of other stars in a galaxy, making it relatively easy to detect even across the vast distances of intergalactic space. Astronomy is one of the few sciences in which non-professionals still regularly make important discoveries and many extragalactic supernovae are discovered by the army of amateur astronomers who scan the heavens from their backyards every night, providing a far more comprehensive surveillance of the sky than could be achieved by the relatively small number of professional observatories. Each new supernova is catalogued, and careful studies are made of the way its brightness flares, peaks and declines, using wavelengths from across the electromagnetic spectrum, from radio waves to gamma rays.

But the details of the explosion and its aftermath are impossible to see over such huge distances and so astronomers are on permanent alert in the hope that the next one might take place within the Milky Way itself. This would not be the first time a local supernova has been detected. Ancient Chinese astronomers were particularly diligent in noting anything new in the heavens, and their records include references to around 20 short-lived 'guest stars' which could be supernovae, with the earliest dating to AD 185. Other notable historic sightings include the supernova of 1054, which was visible in daylight for 23 days and gave rise to the Crab Nebula (*see* Made of Starstuff, page 6), and that of 1572, which was studied by the Danish astronomer Tycho Brahe. The last time a supernova was definitely seen to occur within the Milky Way was over 400 years ago in 1604. At this time in Europe, the prevailing wisdom was that the heavens beyond the Moon and planets were unchanging, so transient objects such as comets and 'new' stars were considered to be phenomena of the Earth's atmosphere. In 1572, Brahe's careful observations had shown that the temporary star did not change its position among the fixed constellations for as long as it was visible, so it must therefore have been very distant. Galileo Galilei made a similar observation in 1604, and used it to argue that the heaven of the fixed stars was not unchanging after all – another piece of evidence that helped to demolish the traditional view of an Earth-centred cosmos.

It is not entirely surprising that supernovae in our galaxy are hard to see. We are deeply embedded within the Milky Way and surrounded by its obscuring clouds of dust, so it is impossible for us to get a complete view of the entire galaxy, at least using light in the visible part of the spectrum. Modern radio, infrared, X-ray and gamma ray telescopes are able to see through this dust, and we also now have neutrino detectors, which can pick up the burst of particles produced by a supernova. But we have only had these capabilities for a handful of decades, so there is plenty of scope for a supernova to have gone off somewhere in the Milky Way during the last few centuries without it being noticed. Indeed, radio and X-ray telescopes have identified clouds of expanding debris from supernovae that occurred over the last few hundred years but were hidden from view by dust.

The only nearby supernova explosion to be directly observed in modern times occurred not in our galaxy itself but around 168,000 light years away in the Large Magellanic Cloud, one of the Milky Way's satellite galaxies. On 24 February 1987, at the edge of a vast region of gas and stars known as the Tarantula Nebula, a previously unremarkable star known as Sanduleak -69° 202 suddenly blazed into prominence, transforming itself into Supernova 1987A, or SN1987A for short. For several months, it was visible to the naked eye in the skies of the southern hemisphere, rapidly becoming the most intensively studied stellar explosion in history.

Of course, the explosion had actually occurred 168,000 years earlier, and the light had been travelling towards us ever since. But three hours before the light of the supernova was picked up by professional and amateur astronomers news of the explosion had already arrived in the form of a burst of neutrinos, tiny particles produced in the supernova's first few moments as the core of Sanduleak -69° 202 collapsed. Although neutrinos travel at slightly less than the speed of light, they had had a head start over the blaze of electromagnetic radiation generated by the explosion. Photons of light are blocked by the dense plasma of superheated gas inside the dying star and are only released when the explosion has blasted the outer layers aside. By contrast, neutrinos barely interact with other forms of matter and so they race outwards from the centre of the star, unimpeded. Their fleeting nature makes them hard to detect, however: although a colossal 99 per cent of the total energy of the supernova is emitted in the form of neutrinos, only 24 of them were actually trapped by detectors here on Earth. But this was far above the normal background neutrino level and a sure sign that something drastic had occurred.

As the initial glare faded away, astronomers were able to follow the aftermath of the supernova in detail. Within weeks, light from the explosion illuminated rings of gas that had been ejected by Sanduleak -69° 202 in the centuries before its core finally collapsed, revealing the star's drawn-out death throes. Years later, the expanding wreckage of the star ploughed into this earlier material, generating shockwaves and causing it to light up once again. As the debris continues to expand into the neighbouring Tarantula Nebula, perhaps we will see at first hand the same processes that triggered the formation of our own Solar System 4.5 billion years ago.

The debris from SN1987A will continue to be studied for centuries to come but it is hard to predict when it might lose its status as the closest supernova explosion of modern times. One candidate with a chance to steal the crown is Betelgeuse, the bright red star that marks the shoulder of Orion. Betelgeuse is a red supergiant star, 1,000 times larger and 100,000 times more luminous than the Sun, and it is very close to the end of its life. Astronomers expect that it could explode at any moment, although in astronomical terms 'any moment' actually means 'any time in the next 100,000 years'. When it does, it will outshine the full Moon and be visible during the day for several months before it fades away, leaving the constellation of Orion changed forever. In fact, as Betelgeuse is 640 light years away it may already have exploded centuries ago – but if so the news has yet to reach us.

Another nearby supernova prospect is the obscure object IK Pegasi, which, at a distance of only 150 light years, is almost too close for comfort. Unlike Betelgeuse, IK Pegasi is not one but two stars: a binary pair in which an ordinary star and a white dwarf circle each other in a close orbit. As is common for supernovae, Betelgeuse will explode when it tries to fuse the iron in its core – an event known as a Type II supernova explosion – but in IK Pegasi the mechanism for the explosion will be rather different. Here, hydrogen gas is being pulled from the normal star by the gravity of its companion and deposited onto the surface of the white dwarf. If the amount of stolen hydrogen reaches a critical level, it will spontaneously undergo a runaway burst of nuclear fusion, destroying the white dwarf in a Type Ia supernova explosion. IK Pegasi is currently barely visible to the naked eye, but if it ever detonates it will briefly become the brightest object in the sky apart from the Sun – although is just far enough away not to pose a direct threat to Earth.

One relatively low-key astronomical development will be the continuing exchange of angular momentum between the Earth and the Moon, which is causing the Earth's spin to slow down while the Moon drifts ever further away at a rate of 4 centimetres every year (*see* Inconstant Moon, page 153). It will be a very gradual process, but in about 250 million years' time the rotation of the Earth will have slowed by an additional 1.5 hours, so that a day will last for 25.5 hours. The Moon is the main driver of the ocean tides and it also helps to stabilize the Earth's axial tilt, keeping the variations that drive the Milankovitch cycles to a minimum. But as it drifts away, the Moon's grip on both the oceans and the Earth's axis will loosen, taking us towards a distant future of tide-less seas and extreme climate fluctuations. But this process will take billions of years to play out to its conclusion and, long before we reach it, other circumstances will affect the oceans, the climate and the planet itself in ways that will make it utterly irrelevant.

The Moon is not the only satellite in the Solar System whose orbit is changing – and some moons are even undergoing the opposite process, as tidal interactions drag them ever closer to their parent planets. In a few million years, Mars' moon Phobos will be just 7,000 kilometres from Mars itself, at which point tidal forces will dismember it into a ring of orbiting debris, which will eventually crash down onto the Martian surface piece by piece. Neptune's large moon Triton and the Jovian satellites Metis and Adrastea will also suffer similar fates at some point in the next few billion years.

Tidal effects on the orbits of moons are relatively simple to predict, but other objects in the Solar System are likely to change their orbits in ways that are harder to anticipate with any certainty. As we have seen, gravitational interactions with the other planets cause the shape of the Earth's orbit to vary, affecting our planet's climate, but we know from studying the planetary systems around other stars that far more drastic orbital changes are possible. In the 1990s, the discovery of 'Hot Jupiters' – giant planets like Jupiter that orbit their stars more closely than Mercury orbits the Sun – was a huge surprise to astronomers because such giant planets can only form where it is cold, far from their parent star. Finding planets this large so close to their star means that they must have moved there after they formed (*see* On Stranger Tides, page 190). This process of planetary migration is now thought to be fairly common in young solar systems – including our own – and is due to powerful gravitational interactions with the remaining gas and rocky debris around the star and to interactions between the newly formed planets themselves.

Although there are obviously no Hot Jupiters orbiting the Sun, it is possible that the two outer planets, Uranus and Neptune, underwent a radical orbital reorganization about half a billion years after the Solar System formed, driven by the combined gravitational forces of Jupiter and Saturn. Computer simulations show how the two planets could have been forced to migrate outwards, moving further from the Sun and even swapping places with each other before settling into their current positions. In the process, they would have disturbed the orbits of millions of smaller objects in the Kuiper Belt, sending a shower of comets towards the worlds of the inner Solar System. Evidence for this planetary rearrangement might be staring us in the face in the form of the Moon's craters. Samples of Moon rock returned by NASA's Apollo missions seem to show that many of the most violent impacts in the history of our satellite occurred between 4.1 and 3.8 billion years ago – exactly the period when Uranus and Neptune are suspected to have been sowing havoc in the Kuiper Belt.

Although the Solar System has settled down since the turbulent days of its youth, we still can't rule out the possibility of further orbital chaos in the future. Gravitational interactions between the planets are relatively weak but over periods of hundreds of millions of years they can add up in unpredictable ways. Computer simulations can help us to explore the range of possible outcomes by fast-forwarding the planetary motions again and again to see how they turn out, and it seems that Mercury, the smallest and innermost of the planets, may be the one that is most at risk of a chaotic change. If so, it will once again be Jupiter that is the main driver for the process as the giant planet's gravity gives Mercury regular tugs, gradually nudging it away from its current orbit. The simulations show that over hundreds of millions of years there is a probability of a small percentage that Mercury could be ejected entirely from the Solar System or plunged into the Sun. There is even a slim chance that it could be propelled onto a collision course with either Venus or – alarmingly – the Earth.

Such a collision between the Earth and Mercury may be very unlikely but there is no doubt at all that our planet will be struck by asteroids and comets many times in the future. As we saw in Small Worlds, there are millions of these objects orbiting the Sun and, although astronomers are making great progress in mapping them and charting their orbits, they are hugely susceptible to the complex gravitational influences of the planets as they move through the Solar System, so it is almost impossible to predict their paths with any accuracy beyond a few hundred years into the future. The impact of an object more than a few kilometres in diameter could have devastating consequences for the Earth's biosphere, leading to a mass extinction of life and a collapse of human civilization. Luckily, this is one disaster scenario over which we may be able to exert some control – space missions such as NASA's Dawn and the European Space Agency's Rosetta craft are providing us with invaluable insights into the nature of these objects, allowing us to envisage ways in which an incoming object could be destroyed or diverted safely away from the Earth. However, due to the inherently unpredictable nature of their orbits, humanity will need to remain vigilant as long as we remain here on Earth.

A collision of an even more spectacular kind is also on the cards – and one that will utterly reshape the appearance of the heavens. Appearing in the night sky as a tiny smudge of light to the side of the Milky Way, Messier 31 – otherwise known as the Andromeda Galaxy – is one of the most remote objects that it is possible to see with the naked eye. Telescopes reveal Andromeda in its full glory: a giant spiral galaxy, even larger than our own Milky Way, with swirling spiral arms studded with brilliant star clusters and clouds of glowing hydrogen gas, shot through with dark lanes of dust. Since this galaxy lies at a distance of about 2.5 million light years from our own, our view of it is also an ancient one: by the time Andromeda's light strikes the retina of your eye it has been travelling for 2.5 million years, so we are seeing the galaxy as it was in the distant past, long before our species even existed. If we wanted to see what Andromeda looks like today, we would have to wait 2.5 million years for the light setting off now to reach us.

However, we won't always have to wait quite so long for news to reach us from our galactic neighbour. In 2012, a Hubble Space Telescope study confirmed not only that Andromeda and the Milky Way are getting closer together – at a relative speed of 110 kilometres per second –but that they are set on a collision course, with the smash due to occur in around 4 billion years' time. The view will be spectacular as the disc of Andromeda, with its hundreds of millions of stars, scythes through the band of the Milky Way. Despite the combined number of stars in the two galaxies – probably over a trillion in total – it is vanishingly unlikely that any of them will actually collide directly during the encounter. On the scale of galaxies, individual stars are simply too small, and the distances between them too large, for head-on stellar collisions to occur. An idea of the sheer unlikelihood of this can be gained by imagining that a marble in the centre of London represents the Sun: the nearest star, Proxima Centauri, would be another marble somewhere in Paris. But, despite a lack of collisions, the gravity of the two galaxies will play havoc with the orbits of individual stars, throwing the neat spiral structures of the two galaxies into chaos. Computer simulations suggest that during this process there is around a 10 per cent chance that the Sun, with all of its remaining planets in tow, will be flung out of the Milky Way entirely to wander the emptiness of inter-galactic space for billions of years to come.

Although there will be no smashes between individual stars, other components of Andromeda and the Milky Way will certainly collide with each other in a dramatic fashion. The great clouds of gas and dust that thread the spiral arms of both galaxies will slam together, creating shockwaves that compress the gas and trigger a huge burst of star formation. Both galaxies will light up with clumps of newborn stars, although the dazzling view will come with its own downside. Many of these new stars will be extremely massive and short-lived and within a few million years the surge of star birth will be followed by a violent burst of star death as supernovae begin to detonate throughout the two galaxies like a volley of firecrackers.

Further fireworks will result from the collision of the supermassive black holes that reside at the heart of each galaxy. Each containing millions of times as much material as a typical star, these two objects are the most massive players in the galactic encounter and gravitational interactions will cause them to spiral in towards each other until they finally merge together, releasing a huge burst of gravitational waves in the process. Astronomers are currently trying to perfect gravitational wave detectors – instruments that can detect the oscillating stretching and squeezing of space-time caused by the acceleration of extremely massive objects – and if they succeed, they should be able to detect the gravitational signature of merging supermassive black holes in other colliding galaxies, giving us a glimpse of the event that will one day occur in our own neighbourhood.

Currently, the supermassive black holes in both Andromeda and the Milky Way are relatively quiet objects. They long ago devoured most of the material within reach of their powerful gravity and now they live at the centres of their respective galaxies on a starvation diet, only occasionally managing to pull in a star or a gas cloud that strays too close. But during the chaos of a galactic collision, it is likely that large amounts of gas from further out in the galaxies' spiral arms could be funnelled to within reach of the newly merged black holes. As it swirls in towards its doom, the gas will heat up, blazing with radiation across the electromagnetic spectrum from radio waves to gamma rays, before disappearing forever into the black hole itself.

For a few tens of thousands of years, the region around the black hole will shine more brightly than all the stars in the two galaxies combined, until all of the gas is swallowed up and the black hole settles back into a quiet existence. We see such objects – known as quasars – in the centres of other colliding galaxies, and often they are so bright that they drown out the light of the surrounding stars. But most of the quasars that we have detected are billions of light years away from us, so we are seeing a phenomenon that was much more common in the distant past than it is today. Astronomers think this is because collisions between galaxies were much more common in the past, since galaxies were still in the process of forming and – in a younger, smaller universe – they were also closer together. Major galaxy collisions, such as that between Andromeda and the Milky Way, will become increasingly rare, and if this one does trigger an episode of quasar activity, it will probably be among the last events of its kind in our corner of the universe.

Eventually things will settle down: like two coalescing swarms of bees, Andromeda and the Milky Way will merge together and where there used to be two spiral galaxies there will now be one large and rather featureless elliptical galaxy, with a single super-massive black hole at its heart. Most of the gas and dust that once enriched the spiral arms of the two progenitor galaxies will either have been converted into stars, swallowed by the super-massive black hole or blasted away into intergalactic space by the radiation from young stars, supernovae and quasar activity. With little remaining gas, the new galaxy will be aging from the start: few new stars will be created from now on.

Andromeda and the Milky Way are the two largest members in the Local Group of galaxies and over the billions of years to come many of the remaining members of this galactic family, including the beautiful Triangulum Galaxy and dozens of smaller dwarf galaxies, will probably also succumb to the combined gravitational pull and be absorbed in their turn. If it survives these merging events without being ejected, our Solar System will eventually be part of a vast but slowly dying galactic giant.

However, while all this is going on the greatest danger to Earth will come not from colliding galaxies or marauding planets but from the very heart of the Solar System itself.

The Sun has nurtured life on our planet for billions of years: as we saw in Sunshine (page 113), its warmth has kept the Earth's surface at just the right temperature to support liquid water and to sustain the complex chemical processes on which life depends and, as the energy source for photosynthesis in plants, sunlight is also the foundation for most of the food chains on which our planet's diverse web of life is based. But the Sun will not always be such a benevolent parent. Deep in the heart of our star is a ticking time bomb that will one day lead to disaster for the Earth and its cargo of living organisms. In the next chapter we will discover what the Sun has store for us as it reaches the end of its life and then look ahead to the far future of the universe itself.

Far Futures

The lifecycle of a star like the Sun follows a predictable path, governed by the physics of gravity, thermodynamics and the nuclear reactions at its centre. After a short-lived and turbulent youth, the Sun is currently in a long, relatively stable phase of the stellar lifecycle known as the main sequence. Deep within its core, it is steadily fusing together the nuclei of hydrogen atoms to make helium and, in the process, generating the heat and light that allow it to shine. It has been doing this for the last 4.5 billion years and will continue to do so for at least another 5 billion years. However, although the Sun may be relatively stable, it is by no means unchanging. As it fuses hydrogen into helium, the concentration of hydrogen in its core gradually decreases and this causes the core to contract, raising the internal temperature and increasing the fusion rate of the remaining hydrogen. This means that as it ages the Sun is using up its hydrogen fuel more and more quickly – and generating more and more energy as it does so. Like all stars on the main sequence, the Sun has been growing steadily hotter and brighter throughout its life, increasing its output of heat and light by about 10 per cent every billion years. This brightening will inevitably continue into the future, and eventually the warmth that first made the Earth hospitable to life will instead begin to make conditions on our planet more and more uncomfortable.

For the last few billion years, Earth has managed to remain at a roughly constant temperature because its atmosphere and rocks have acted like a natural thermostat. As the Sun's energy output has increased, this has accelerated the rate at which silicate rocks on the Earth's surface are weathered away, a process that also removes carbon dioxide from the atmosphere. Since carbon dioxide is a greenhouse gas, its removal means that the atmosphere retains less of the Sun's heat and so the increase in incoming solar radiation is balanced out. The carbon dioxide is eventually returned to the atmosphere through volcanic processes but as long as CO_2 is removed faster than it is returned the overall atmospheric level will continue to fall and temperatures will remain roughly stable.

This negative feedback cycle, in which an increase in solar radiation leads to a decrease in the amount of atmospheric carbon dioxide, has served our planet well for billions of years, maintaining the temperature conditions under which water is a liquid and life can flourish. But it also has a negative side. As well as helping to regulate our planet's temperature, carbon dioxide is vital for the process of photosynthesis, by which plants transform the Sun's energy into food. As carbon dioxide levels continue to drop, photosynthesis – and with it the basis for most of the Earth's food chains – is heading for a disaster.

Scientists predict that the first critical point will be reached in about 600 million years from now. When the amount of carbon dioxide in the atmosphere falls to below 50 parts per million (compared with around 400 parts per million today) around 99 per cent of current plant species will be unable to photosynthesize, despite the abundance of sunlight pouring down on them from the brightening Sun. The remaining 1 per cent of plant species – many of them grasses – use a slightly different chemical pathway during photosynthesis, and can continue to function with much lower levels of carbon dioxide. These survivors may be able to continue for another several hundred million years, presumably evolving to fill the ecological niches vacated by their less fortunate relatives. But eventually, they too will succumb to a lack of carbon dioxide, and the Earth's era of lush forests and verdant grasslands will be at an end. With the plants will go the Earth's oxygen-rich atmosphere and ozone layer – also by-products of photosynthesis – and without plants to eat, oxygen to breathe or a barrier against the Sun's harsh ultraviolet rays, the animal population will rapidly follow the plants into extinction. Of course, evolution may once again come into play and it is possible that the final extinction of complex plants and animals may be kept at bay for a while by ever more ingenious adaptations to the worsening conditions. But in the end, the Sun will inevitably win the battle and the future Earth will be inhabited only by simple microbes, just as it was for the first few billion years of its history.

A catastrophe caused by too little carbon dioxide in the atmosphere might seem ironic given current concerns about global warming caused by the opposite problem of too much carbon dioxide. Could it even be that by burning fossil fuels we are helping to stave off this ultimate environmental disaster? Unfortunately, the answer is a resounding no. By burning coal and oil we are effectively resurrecting carbon dioxide that was extracted from the atmosphere by plants and other organisms hundreds of millions of years ago and then buried underground when they were fossilized. The liberated carbon dioxide has the potential to raise global temperatures and alter the Earth's climate for thousands of years to come – long enough to cause serious problems for many generations of our descendants – but this is only tinkering around the edges of the vast geological processes that shape the Earth over timescales of millions of years. In the long term, our carbon emissions will be reabsorbed and our reckless experiment with CO_2 will become another blip in the geological history of the Earth – albeit a damaging one for us and the living things with which we currently share the planet. We should not flatter ourselves that, just because we are capable of despoiling our own environment in the geological short term, we have the power to save the planet from its ultimate fate.

As the Sun continues to brighten and the weathering of rocks accelerates, eventually there will be no more carbon dioxide to remove: the thermostat will fail completely and the planet's temperature will inexorably start to rise. In something over a billion years' time, it is likely that temperatures will reach the point when the oceans will begin to evaporate. The atmosphere will fill with water vapour and the Earth will enter a new phase in its history. Like carbon dioxide, water vapour is a potent greenhouse gas and as the increasing solar brightness liberates more and more of it into the air this in turn will trap more and more heat in the atmosphere. Instead of negative feedback, now the feedback will be positive: higher temperatures mean more water vapour, which traps more heat and raises the temperature still further – and so the cycle will continue. Instead of a self-regulating thermostat, the dominant process will now be a runaway greenhouse effect, driven by evaporation. As the oceans dwindle away and temperatures soar, even microbial life will find conditions increasingly difficult. Water – the alien substance whose arrival on Earth billions of years ago helped to make our planet hospitable to life – will become complicit in life's ultimate extinction. This may take some time, however. As the continents revert to deserts and the ocean basins are transformed into vast saltpans, some microbes could retreat underground where pockets of water might remain. Here the Earth's final inhabitants could conceivably persist for a billion years or more – a strange underground end to life's long existence on our planet.

The loss of the oceans might also have another consequence. Without water to soften the rocks and lubricate the motion of the plates that make up the Earth's crust, plate tectonics will grind to a halt. The effects of this could reach to the very core of the planet, disrupting the turbulent flows of material that transport heat from the centre to the surface of the Earth, and eventually causing the liquid iron dynamo of the outer core to shut down. Without the dynamo, the Earth's magnetic field will fail and the planet's atmosphere will be exposed to the full force of the solar wind. Even if the magnetic field somehow survives the end of plate tectonics, it is ultimately doomed: in 3 to 4 billion years' time the liquid outer core will have completed its long process of solidification and the Earth's internal dynamo will be gone forever.

There is a silver lining to the Sun's inexorable brightening: its Habitable Zone – the 'Goldilocks' region of space in which temperatures are just right for water to remain liquid – will gradually move further out, leaving the Earth behind but perhaps bringing balmier conditions to Mars and the icy moons of Jupiter. But even if our descendants follow the Habitable Zone as it expands to include these more distant worlds, this will only be a temporary reprieve. In around 5 billion years' time, the Sun will reach the end of its career as a main sequence star, when all of the hydrogen in its core has finally been used up. At this point, the core, now composed almost entirely of helium 'ash', will shrink still further, raising the temperature until a new bout of hydrogen fusion is triggered in a shell surrounding the core itself. Destabilized by this new source of energy, the outer layers of the Sun will begin to expand and, over the next 2 billion years, our star will gradually swell to more than 100 times its current diameter. As it does so, the Sun's surface will cool to around 2,300 degrees Celsius, going from yellow-white to red in colour, but its overall luminosity will increase by a factor of up to 2,700 times, flooding the Solar System with light and heat. Our star will have become a red giant.

This colossal increase in size won't be without consequences for the planets in the Solar System. Mercury and Venus will be swallowed up and destroyed and it seems likely that the Earth will share their fate, as the Sun's extended atmosphere exerts a tidal drag on our planet's orbital motion, causing it to spiral inwards. But long before our planet meets its fiery end, a combination of the heat radiating from the bloated Sun and a huge increase in the solar wind will have stripped the Earth of its atmosphere and blasted the surface, reducing it to a smouldering cinder.

Once again, there may be a silver lining, as the increased heat and light shift the Habitable Zone out still further, to embrace

Saturn's large moon Titan. But the Sun's evolution will not end here and it will prove to be a fickle benefactor. As fusion continues in the shell of hydrogen around the core, new deposits of helium ash will build up until the core itself reaches a critical mass, triggering a 'helium flash' in which the core begins to fuse helium into carbon. In a reversal of its previous growth, the Sun's outer layers will contract and the star will shrink by a factor of about 20. As it does, its luminosity will decrease and the Habitable Zone will retreat inwards, abandoning the outer moons and planets of the Solar System to the cold of space. But deep inside the Sun, this new equilibrium will be short-lived: after only 100 million years or so, the core's reserves of helium will be exhausted and again it will resort to the fusion of hydrogen in a shell surrounding the core. For a second time, the Sun's outer layers will expand, but this really will be the end of the line: there will only be enough fuel to sustain fusion for another 30 million years.

As its fusion reactions falter, the Sun will shed its outer layers, blowing them gently into space to create a vast halo of gas up to a light year across. This huge reduction in the Sun's mass could severely disrupt the orbits of any remaining planets, spelling the end of the Solar System as we know it. The process will take just a few tens of thousands of years and will leave behind the Sun's exposed core in the form of a white dwarf star – an intensely hot sphere of carbon and oxygen about the size of the Earth, but containing more than 100,000 times as much material. The fierce radiation from the white dwarf will illuminate the surrounding gas, causing it to glow and creating a planetary nebula. We see many examples of these planetary nebulae in the

Milky Way today – relics of Sun-like stars that have already reached the end of their lives.

Planetary nebulae are some of the most beautiful objects in the night sky but their name is somewhat misleading since these stellar tombstones really have very little to do with planets. It was coined in the eighteenth century by astronomer William Herschel since, through telescopes of the time, they appeared as fuzzy, vaguely round blobs that reminded him of planets. However, modern telescopes reveal them in glorious detail, with shells and streamers of glowing gas often twisted into elaborate cones and spirals by the rotation of the dying star. If nothing else, the end of the Solar System will be very beautiful. As the gases of the planetary nebula dissipate into space, they will carry with them the helium and carbon that were created during the Sun's 12-billion-year career as a fusion factory. Eventually, some of this material will go on to form new stars, with their own planets – and perhaps even life.

Meanwhile, the dense core of the Sun will still have a long afterlife ahead of it. White dwarf stars are supported against their own gravity not by nuclear fusion but by the jostling of closely packed electrons. With no ongoing fusion reactions to generate energy, the white dwarf will inexorably grow cooler and dimmer, like a flashlight with a fading battery, but the process will be extremely slow, taking many billions of years – far longer than the time the Sun spent as a life-giving main sequence star. Gradually, the carbon and oxygen will crystallize and the Sun will end its career as a cold, dark hulk, orbited by the frozen remains of the outer planets.

The Sun's fate was set from the moment it formed, since all the different stages of its life, their order and duration, are all determined by its initial mass and composition – the amount of material it contains and the ratio of different elements in its makeup. Like that of the Sun, the ultimate fate of the universe will also be determined by its contents – the matter and energy that were created in the fury of the Big Bang – so its future too has been set by the circumstances of its birth. However, although we now have an excellent understanding of the principles of stellar physics, our understanding of the earliest moments of creation is still imperfect and even our knowledge of the current content of the universe remains rather patchy. The destiny of the cosmos therefore remains an area of active investigation and debate – and there is plenty of room for speculation. However, based on what we know of the last 13.8 billion years of cosmic history, it is possible to extrapolate current trends for billions or even trillions of years into the future.

Today, the universe is still expanding – the legacy of its explosive origin – but as time goes on we might expect gravity to work against the expansion as the mutual attraction of all the matter in the cosmos – planets, gas, stars and galaxies – acts to pull it all back together. For much of the twentieth century, it seemed that the future of the universe would be determined simply by how much matter it contained and three possible scenarios suggested themselves. With too little matter, gravity would slow the expansion but never quite halt it. In this case, the universe would be 'open' – continuing to grow forever and becoming progressively darker and emptier as time went on. With significantly more matter, the universe would be 'closed':

gravity would eventually halt and then reverse the expansion, and everything would collapse back in on itself, towards an incandescent 'Big Crunch' – and perhaps triggering a new Big Bang to set things off all over again. Finally, with precisely the right amount of matter, the universe could be poised between these two possibilities, with the expansion coasting towards a gentle halt at an infinitely distant time in the future.

How much matter is 'just right' is a difficult question to answer, but first of all, the quest to determine how much matter the universe contains has not been a straightforward one. Since the 1930s, astronomers have been finding evidence that the universe contains far more matter than we can actually see, even with our battery of telescopes and cameras to detect emissions from beyond the visible part of the spectrum. In particular, the movements of stars within galaxies, and of individual galaxies within galaxy clusters, are too rapid to be explained merely by the gravity of the visible stars, gas and dust – and even exotic but hard-to-see objects such as black holes cannot possibly exist in sufficient numbers to provide the additional gravity required. The problem was sidelined for several decades until, in the 1970s, the pioneering work of American astronomer Vera Rubin forced the astronomical community to face up to a disconcerting prospect: a large fraction of the universe might be made up of something that they couldn't see. Since then, a body of evidence has confirmed the presence in the universe of large quantities of mysterious matter that neither emits, reflects or absorbs light – 'dark matter' (see And Yet It Moves, page 218). There is a lot of this mysterious stuff around: about five times as much dark matter as ordinary matter. But is this enough to halt the expansion of the universe?

Scientists are trained to be objective and to put the evidence about how the universe actually is before their personal preferences about how they would like it to be. However, when it comes to a subject as profound as the fate of the universe, everyone naturally has their own feelings about what they would like the answer to be. In her poem 'Let There Always Be Light (searching for dark matter)', the astronomer and poet Rebecca Elson made it clear where her preference lay, writing 'Let there be enough to bring it back / From its own edges'. Despite the alarming prospect of a future Big Crunch, many of us would probably agree with Elson that it is in some ways a more appealing scenario than that of a universe that simply grows inexorably larger, darker and emptier. Other astronomers argued from philosophical principles that the universe was probably perfectly balanced, forever slowing towards a halt but never quite reaching the point of collapse. But everyone had reckoned without a further discovery, which would turn all their theories upside down.

In 1998, an international team of astronomers made a shocking announcement. They had been observing supernova explosions in distant galaxies, but something about their results was refusing to add up. The most distant supernovae were consistently fainter than would be expected if the expansion of the universe had simply been slowing down under the influence of gravity. In fact, the most plausible explanation for the faintness of the supernovae was that, for the last 5 billion years or so, the expansion of the universe has actually been *speeding up*.

It seemed that something had been missing all along from our picture of the universe – a mysterious form of energy that pervades all of space and works against gravity, pushing the universe apart at an ever-increasing rate. Since the nature of this energy remains unknown, scientists have dubbed it 'dark energy' and astonishingly it makes up around 68 per cent of all the 'stuff' in the universe (by contrast, dark matter makes up 27 per cent and ordinary matter only 5 per cent).

Where matter is abundant – in groups and clusters of galaxies – its mutual gravity is enough to hold these structures together against the cosmic expansion of space. But in the vast voids between the galaxy clusters, dark energy rules. As space expands, the clusters become ever more isolated by distance, and the gravity between them weakens. But dark energy is a property of space itself and so, as space expands, dark energy's influence also grows – and the expansion rate therefore increases with time.

Once astronomers knew what to look for, they began to find the telltale signatures of dark energy in other places. As with dark matter, there are now several independent lines of evidence – from the Cosmic Microwave Background radiation to the large-scale distribution galaxy clusters – which all point to dark energy's presence. Even more than dark matter, it has dominated the evolution of the universe for the last 5 billion years and will clearly play a defining role in how the universe evolves in the future. But here we have a problem: because we don't understand what dark energy is, or why it has behaved the way it has in the past, it is very hard to say for certain how it will continue to affect the universe in the future – although naturally cosmologists already have many theories, and the race is now on to find evidence to either refute or confirm them.

The simplest assumption might be that things go on in much the same way as they are at present – in which case the universe will continue to expand more and more rapidly and the distances between structures that are not gravitationally bound to each other will become ever greater. Galaxy clusters will be carried away from one another like patches of foam on an ever-widening sea. Surprisingly, this doesn't contradict Einstein's rule that nothing can travel faster than the speed of light. When visualizing the expanding universe, it is tempting to imagine galaxies being flung apart and speeding away into a pre-existing empty void, but in fact they are technically not moving at all: the space between them is simply getting bigger. Over large distances, the expansion can be so great that the distance is growing larger more quickly than light can traverse it – but the galaxies at either end are not moving through space, so no laws of physics are being broken.

As time goes on, we will therefore see the most remote galaxies fade from our view and no communication or interaction with them will ever be possible again. In this way, most of the universe will eventually become inaccessible to us. First distant and then nearby galaxy superclusters will be carried away from us and lost from sight, until the visible universe consists only of the remaining galaxies of our own Local Group, held together by their mutual gravity.

In an extreme version of this scenario, if dark energy's influence continues to increase indefinitely, the universe could be headed for what cosmologists have called the 'Big Rip'. As the rate of expansion accelerates, eventually it will start to gnaw away at the galaxy clusters themselves, overwhelming their gravitational bonds and isolating individual galaxies into their own regions of expanding space. Galaxies will in turn succumb: their stars will be separated, solar systems will be dismembered, and stars and planets will be ripped apart. Eventually, dark energy will even swamp the electromagnetic forces that hold molecules together, stripping them down to their constituent atoms, and then overcoming the nuclear forces that bind the nuclei of atoms themselves. The entire universe will be reduced to its smallest constituents, each irrevocably isolated from all the others by distances approaching the infinite. By some reckonings, this process could occur as soon as 20 billion years' time – which would mean that the universe is already more than a third of the way through its life.

However, our understanding of dark energy is still far too basic for us to foretell its behaviour with any degree of certainty. Other hypotheses suggest that its influence might even reverse, halting the expansion and forcing the universe back towards a Big Crunch. In terms of predicting the ultimate fate of the universe, the possibilities are still wide open.

Assuming the cosmos doesn't end in a Big Rip or a Big Crunch any time soon, our galaxy could continue to change and evolve long after the last stars have ceased to shine in about 120 trillion years' time (a trillion is a million times a million, so this would be almost 10,000 times the current age of the universe). By this stage, it will consist of a swarm of stellar corpses – white dwarfs, neutron stars and black holes – all orbiting around the galaxy's central supermassive black hole. With all of the gas and dust either already converted into stars or expelled from the galaxy, no new stars will form, but the occasional collision of binary pairs of dead stars will provide brief illumination in the form of a supernova explosion. Over timescales of quintillions (millions of millions of millions) of years, gravitational encounters between the dead stars will fling some of them out of the galaxy entirely and send others tumbling to join the supermassive black hole at its centre.

Beyond this stupendously remote era, the picture becomes extremely hazy and the evolution of the universe depends increasingly on aspects of physics that we do not completely understand. Never mind the mysteries of dark energy and dark matter, over such vast timescales even the behaviour of ordinary matter starts to become mysterious. For example, if, as suggested by some versions of particle physics, the protons in atomic nuclei are ultimately unstable, then eventually even white dwarfs and neutron stars will decay into radiation and an array of smaller fundamental particles.

Black holes will then be the only surviving remnants of our own era of stars and galaxies. But even black holes are not eternal. The work of Stephen Hawking and other physicists predicts that the uncertainty principle, one of the stranger aspects of quantum mechanics, should cause black holes to 'evaporate' over very long timescales, dissipating themselves in a stream of fundamental particles. If the universe persists indefinitely, then it will come to be dominated by such weird physics. Quantum events that have been vanishingly improbable over the current lifetime of the universe will have plenty of time in which to occur again and again during the trillions upon trillions of years of future time, and matter itself will become a slippery concept as particles morph and change or flit in and out of existence. Given enough time, even an event as unlikely as a new universe springing spontaneously into being could become a near certainty.

It might seem as if we have come full circle, with the predictions of astronomy becoming as mysterious and uncertain as those of astrology. But scientists are continuing to use observation, experiment and reasoning to increase our understanding of how the universe works – and hence to glimpse what it has in store for us.

Our current, imperfect view of the future gives us plenty to think about, even if we are currently unable to say for sure what the ultimate fate of the universe will be. Nothing – not the Earth, the Sun, or even the Milky Way galaxy – will last forever, and it seems that we are living in a long but finite period of cosmic history in which stars, planets and complex carbon-based life are all possible. We have seen how much the universe has changed and evolved in the 13.8 billion years since it emerged from the searing brilliance of the Big Bang and perhaps we should be reassured to know that there will be many billions

– perhaps trillions – of years of equally momentous developments still to come. Even after our Sun and the Earth are gone, there will be new stars and new planets for us to explore for a long time to come – far longer than the history of the universe to date. Given that our species has only been around for 200,000 years or so, and that our current understanding of the cosmos has been achieved within living memory, the habitable future of the universe stretches before us almost like an eternity. We will have plenty of time to decide how to use this knowledge to shape our own future. And if some form of intelligence persists beyond our era of bright stars and cosy planets, into the long ages of darkness and strange physics, it may find more than enough complexity and beauty to keep it occupied for eternities to come.

Space Culture

At one time or another almost everyone has looked up at a night sky full of stars and wondered about their place in the grand scheme of things. The more we learn about our universe the more astonishing and awe-inspiring those starry vistas seem, and the questions that astronomy tries to answer are some of the biggest of all: how did everything begin? When will it end? Are we alone?

These are profound, philosophical ideas but they are not divorced from our everyday lives and the quest to explore and understand the universe beyond the Earth is more than just the pursuit of knowledge for its own sake. On the contrary, as we've seen, objects and events in space have been shaping our planet for billions of years and they continue to influence almost every aspect of the world around us today. Everywhere we look here on Earth, we can find a story that has its beginning out there in space.

For the last few decades astronomy and space science have played an important role in helping us to understand the natural environment here on Earth. In the late 1960s and early 1970s NASA's Apollo missions returned iconic photographs of Earth as seen from our nearest celestial neighbour, the Moon, revealing a fragile blue-and-white globe hanging in the immense blackness of space. At the time these images, startling in their beauty, galvanized the emerging environmental movement and our planet is now continually monitored by dozens of artificial satellites, beaming back streams of images and data on the Earth's land, oceans, atmosphere and living systems. As human pressure on ecosystems and natural resources continues to grow this ability to study our planet as a single interconnected whole has never been more important.

Being able to monitor the Earth from space is vital, but valuable insights into the processes that define our home planet can also come from much further afield. As we've seen, our neighbouring planet Venus, cloaked in a thick blanket of carbon dioxide gas and with surface temperatures of over 450° Celsius, is an object lesson in the dangers of a runaway planetary greenhouse effect, while on Saturn's giant moon Titan rivers and seas of liquid hydrocarbons hold up an intriguing mirror to the watery processes that have shaped Earth. Meanwhile our own Moon's desolate surface holds ancient clues to Earth's geological past, and the rocks of Mars may even provide insights into the origins of life itself.

Astronomical research has an impact on our culture as well as our natural environment. From science programmes on television to full-colour spreads in the national press, the latest discoveries in space are presented and discussed alongside news and politics, sport and art. Pop culture is saturated with the language of astronomy and its imagery appears in movies, video games and advertising – and even high street fashion. Meanwhile contemporary artists share the public's fascination with space, and their responses to astronomical research, and the questions that it poses about our place in the cosmos, are displayed in galleries around the world.

As we've seen, this popularity is nothing new. The revolutionary ideas of Galileo were eagerly consumed by people across Europe, and over the subsequent four centuries astronomy has continued to enjoy a world-wide appeal, gripping the public imagination in ways that many other academic disciplines can only envy. In eighteenth-century London, as astronomers prepared to observe the rare transit of Venus across the face of the Sun, fashionable society from the king to the coffeehouse looked on with excited interest. A century later public lectures by eminent Victorian astronomers were a mainstay of the self-improvement culture of working men's societies. And, as television assumed a central place in twentieth-century life, the mind-blowing concepts and beguiling images of astronomy found their way onto the small screen, where a handful of astronomers even began to carve out a niche as unlikely TV celebrities.

It isn't difficult to understand the reasons behind astronomy's current popularity. You don't have to be a space scientist to feel a tingle of excitement as a robotic rover explores the dusty land-scapes of Mars and beams back images of scenes that no human being has ever glimpsed before. It's not just professional astron-omers who appreciate the thrill of discovery when a brand new planet is found orbiting a distant star. And it would certainly be hard to deny the beauty and power of the unearthly vistas revealed by modern instruments like the Hubble Space Telescope.

But the impact of astronomy and space science isn't just about its ability to provoke a sense of awe and wonder. The quest to understand the cosmos has changed the way we think about ourselves, and many powerful new technologies that were first developed in order to explore the universe have found far-reaching and unexpected uses beyond the walls and domes of the astronomical observatory.

Right from the start, astronomy has always been a practical science. For millennia, before the invention of reliable mechanical clocks, the repeating cycles of the heavens were the only way of keeping an accurate track of the passage of time. The Sun, Moon and stars acted as clock and calendar, marking out hours and days, months and years. This was more than just a convenience: for a farmer, being able to predict the seasons is vital for knowing when to plant your crops or move your livestock to the best pastures, while complex urban societies would quickly descend into chaos if no one could agree on the time or date.

Even today, when atomic clocks can keep time to an accuracy of better than a second in every hundred million years, our biology keeps us tied to the ancient rhythms of days and seasons. We now know that the day–night cycle of the Earth's rotation and the yearly circuit of our orbit around the Sun are not perfect timekeepers. They are subject to long-term trends and random variations, slowing down or speeding up in ways that are not always easy to predict in advance. By international agreement, our super-stable atomic time is frequently adjusted to keep it in line with these erratic changes, allowing our clocks to remain synchronized with the astronomical patterns of night and day, summer and winter.

We've seen how, for our ancestors, the stars also played a vital role in getting around: before accurate maps and satellite navigation devices the most reliable way to find out where on Earth you were was by observing the heavens. Even in today's era of hand-held SatNav units many sailors still choose to learn the tried and tested principles of old fashioned astronavigation. After all, technology isn't infallible and you never know when you might need a star to guide you safely back to port. In fact, modern SatNav is really just a new twist on an old astronomical idea: instead of measuring our position on Earth according to the Sun, Moon and stars, today we do it with reference to a constellation of artificial satellites.

We've also seen how in the sixteenth and seventeenth centuries, as developments in technology and theory brought fresh insights into the workings of the universe, the science of astronomy began to exert a profound influence on society itself, helping to lay the philosophical foundations of the modern world. The work of astronomers such as Copernicus, Kepler and Galileo demolished the long-held view that Earth was at the centre of the universe, challenging the cosy certainty that human beings occupied a privileged place in the cosmos. This momentous shift in perspective helped to set the scene for the scientific advances of the Enlightenment and the Industrial Revolution.

Today, astronomical research continues to provide a steady stream of tangible benefits for our daily lives. Many people know that the World Wide Web had its origins in the particle physics labs of CERN in Geneva, where researchers were looking for new ways to share the huge quantities of data being generated by their experiments. But few are aware that wi-fi, another essential component of the modern internet, relies on technology originally developed in the 1980s by Australian radio astronomers who were hunting for the elusive signatures of black holes. Every day over a billion people now use the invention that this search inspired, as they surf the web from their laptops or mobile devices.

At around the same time, radio astronomers in the UK were also trying to solve a tricky astronomical problem: how to accurately pinpoint the positions of their radio telescopes in order to observe the cores of extremely distant galaxies known as quasars. The technological solution they devised has since found a new home in millions of mobile phones, where it enables the handsets to calculate their location relative to the satellites of the Global Positioning System, even when the signals are weak or distorted.

Astronomy has also played a pivotal role in the development of another technology that we use every day: the camera. When photography was first developed in the first half of the nineteenth century, astronomers were amongst the first to recognize its potential as a scientific tool for recording their observations, and even for capturing light that was invisible to human eyes. But early cameras struggled to detect the faint glow of celestial objects and so astronomers took up the task of improving the technology to the point at which it would be of use to them.

The nineteenth-century astronomer Sir John Herschel was one of these photographic pioneers and several fundamental advances in photography came directly from his experiments. Herschel is credited with the invention of the cyanotype – precursor to the blueprint – and the platinum process for developing photographic prints. Even our language owes a debt to Herschel's work in this field: he was the first to coin the English word 'photography' and to apply the descriptions 'positive' and 'negative' to the photographic process. He even came up with the term 'snapshot'.

This close relationship between astronomy and photography continues to the present day. By the 1970s astronomers were seeking new imaging technologies that could improve on the light sensitivity and level of detail achieved by the best photographic films and emulsions. The result was the development of electronic 'charge-coupled device' (CCD) detectors, which soon became standard camera components for a new generation of powerful instruments such as the Hubble Space Telescope. The descendants of these detector chips can now be found in everyday life, at the heart of mass-market digital cameras.

Anyone in possession of a modern smartphone is therefore carrying no less than three astronomically inspired pieces of technology around in their pockets, in the form of wi-fi, SatNav and digital camera chips. But the practical benefits of astronomy and space science go further than mere convenience – they can even help to keep us healthy.

Technology developed by astronomers to give their telescopes a sharper view of the stars is now used by opticians to diagnose and treat problems in the human eye. Even the largest and most powerful telescopes here on Earth must peer up at the heavens through our planet's atmosphere, and the chaotic movements of this thick blanket of air cause light from celestial objects to blur and distort. This familiar 'twinkling' effect limits the sharpness of astronomical observations and means that large Earth-bound telescopes are prevented from fully resolving the smallest details of which they are capable. To get around this limitation astronomers and engineers now equip modern telescopes with 'adaptive optics' (AO) systems, devices and software that continually monitor the distortions introduced by the atmosphere and then cancel them out by rapidly altering the shape of the telescope's mirror in order to precisely counteract them.

These systems have proved remarkably effective, and many ground-based telescopes now routinely achieve a level of detail comparable to that of the Hubble Space Telescope, which orbits high above the Earth's turbulent air. Like other techniques that were developed to help us explore the universe, adaptive optics has now found other uses here on Earth, this time in medicine, where it is employed by opticians to measure aberrations in the human eye and even in some forms of laser eye surgery.

The world of healthcare has also benefited from software that was originally written by astronomers and computer code devised to answer esoteric questions about the distant reaches of the universe is being pressed into service in hospitals as well as observatories. Programs designed to analyse the huge amounts of data generated by spectroscopic galaxy surveys have recently been used by UK researchers to improve the efficiency of medical scanning procedures. Doctors using these re-purposed algorithms can now perform up to ten times more scans per day, greatly increasing the number of patients seen. Other software designed to recognize the faint smudges of galaxies in grainy photographic images of the sky has now found a use identifying clusters of cancerous cells in biopsy samples, thus helping to diagnose potentially fatal diseases.

Of course, the process by which these benefits are derived is unpredictable and it's certainly very hard to say in advance which line of astronomical research might ultimately lead to a practical application back on Earth. But often in science it's the toughest engineering problems and the trickiest theoretical questions that inspire the most dazzling technological break-throughs. As a field that involves some of the most challenging questions of all, it's perhaps no wonder that astronomy continues to generate a portfolio of useful spinoffs.

By now it should be plain that life on our planet is as much a product of its cosmic environment as it is of local conditions here on Earth. But for the first time in human history the process is no longer exclusively one-way. For almost a century, humanity has been making its mark beyond the Earth, and every year our influence is reaching further and further out into space.

Since the Space Age began, with the launch of the Soviet satellite Sputnik 1 in 1957, access to the space above our heads has increasingly come to assume a central role in politics, economics and the technological infrastructure of daily life. The early triumphs of the 'space race' in the 1960s and 1970s may have owed much to the military rivalries of the Cold War but Yuri Gagarin's first manned orbit of the Earth and Neil Armstrong's 'one small step' on the Moon made them both into global as well as national heroes.

Today, the human presence in space is characterized to a large extent by cooperation rather than rivalry, although this probably has as much to do with a desire to share the financial costs of human spaceflight as it does with ideas of global peace and love. However, perhaps we shouldn't be too cynical about the brave new era of pragmatic space programmes. The International Space Station was constructed by a consortium of bickering nations who can rarely agree on anything here on Earth, and yet in space they have shown themselves capable of working together over a sustained period to create one of the most complex and challenging structures in human history. Indeed, the unsung triumph of the International Space Station is perhaps that it managed to be built at all.

For the time being, the cosmonauts and astronauts on the ISS are our only permanent human representatives in space, but already we have left our footprints on the airless surface of the Moon, where they will remain for millennia until they are gradually erased by meteorite impacts and the occasional moonquake. Future manned missions to the Moon, asteroids and even Mars are on the drawing board and all are likely to be international efforts, but space is no longer the exclusive preserve of powerful governments. The Outer Space Treaty of 1967 sets limits on the commercial and military exploitation of space, but the prospect of space tourism is just around the corner and there is talk of mounting expeditions to mine valuable minerals from the Moon and asteroids.

In the meantime, human technology has travelled even further afield. From the halo of artificial satellites circling our planet, to the fleet of long-range space probes scattered across the Solar

System, and out into the constantly expanding bubble of radio emissions that currently extends among the stars for tens of light years in all directions, we are inexorably spreading our artefacts and our culture into the galaxy.

As well as currently active craft like New Horizons, Dawn and Rosetta our corner of the universe is increasingly populated with the relics of our previous missions and there is even a new field of 'space archaeology' devoted to studying and preserving the history of the Space Age. Right now, dozens of defunct space-craft dot the alien landscapes of the Moon, Venus and Mars. The European Space Agency's Huygens probe lies beneath the orange skies and methane rain of Titan while the Philae lander will accompany Comet 67P/Churyumov–Gerasimenko on its elon-gated journeys around the Sun until the comet finally disspates in a puff of gas and dust. On the 30-kilometre-long space rock Eros, NASA's NEAR/Shoemaker probe rests upon a bed of fine asteroidal dust, while spread throughout Jupiter's vast, cloudy atmosphere is a light sprinkling of alien atoms and molecules – the remains of the Galileo spacecraft that ended its eight-year mission with a fiery plunge into the giant planet in 2003. Many more dead and defunct craft are permanently orbiting the Sun, and closer to home the debris of more than half a century of rocket launches and obsolete satellites now litters Earth orbit as 'space junk', creating a mounting hazard for new space missions.

Technology has also left its imprint in other ways. There are artificial craters on the Moon and comet Tempel 1, created when spacecraft were deliberately crashed into them in order to study their structure and composition. The tracks of buggies and rovers mark the dusty surfaces of the Moon and Mars, and the Curiosity rover has even drilled into various Martian rocks to analyse their chemistry, in the process creating a series of perfectly circular holes that look for all the world like enigmatic works of art. Indeed there is already art on Mars. In January 2015 the UK's Beagle 2 lander was spotted in satellite images of the Martian surface, twelve years after the craft had gone missing. It seems that Beagle 2 had landed safely but its solar panels failed to fully deploy, preventing it from carrying out its mission or contacting home. Though it may not have succeeded in its scientific goals it could still be considered to have achieved an artistic first: on-board the spacecraft is a distinctive spot-painting by the artist Damien Hirst as well as an electronic call-sign composed by the band Blur.

On 25 August 2012 we even became an interstellar species when the unmanned Voyager 2 probe passed through the heliopause, the boundary between the solar wind and interstellar space, after traversing the Solar System for almost 35 years. In the coming decades it will be joined by four more craft as Voyager 2, Pioneer 10 and 11 and New Horizons all follow it across the Solar System's outer edge. Even travelling at several kilometres per second it will be tens of thousands of years before these representatives of our technology pass anywhere near another star, but all of them contain mementos and messages from planet Earth – just in case someone ever does come across them and wonders about the beings who launched them out into the cosmos.

In fact human culture has already crossed the mind-numbing gulfs between the stars, travelling at the speed of light. Since the early twentieth century, radio and television signals and pulses from our radar systems have been trickling inadvertently into space and they now fill a spherical volume around a hundred light years in radius, encompassing hundreds of stars and planetary systems. Whether or not there is anyone out there to listen to them, these electromagnetic emissaries will continue to expand into the cosmos forever, gradually growing fainter until they fade into the hiss of background radiation.

Our presence is no longer confined to a single planet orbiting an unremarkable star. Human artefacts and ideas will probably endure in some form whatever happens to our species here on Earth – but will anybody else ever find them? There is still so much for us to learn about the origins and evolution of life on our own planet, but it seems that the basic ingredients and conditions for biology to get started might be common throughout the universe, and for several decades astronomers have been scanning the skies for signals that might indicate the presence of intelligent beings with a technological civilization. So far we have drawn a blank, but the universe is vast and we have a long way to go before we exhaust all of the possible places in which to look. It seems unlikely that we will give up the search any time soon, simply because the implications of finding that we are not alone in the cosmos would be as profound and far-reaching as anything we have discovered so far. Like Galileo's observations of the moons of Jupiter it would change forever our ideas of who we are and how we relate to the rest of the universe, although it's impossible to predict what the philosophical, technological and social consequences of making contact with alien minds might be. What would we make of a civilization that could easily be thousands or even millions of years more advanced than our own? Perhaps more to the point, what would they make of us?

Even if we ultimately discover that we are all alone in the cosmos, this too should give us pause for thought. Over the last four centuries our telescopes have revealed a universe that is larger and more complicated than anything Galileo could have imagined. It's full of amazing objects and phenomena, from galaxies and nebulae to black holes, neutron stars, dark matter and dark energy, but among all of these astronomical marvels by far the most complex and astonishing thing that we have discovered is right here on Earth: the human brain.

Right now there are seven billion of them, all interacting with each other and with the universe around them, constantly combining and recombining experiences, evidence, ideas and imagination to produce science, history, art, music, architecture and every other expression of what it is to be a human being. Between us, over thousands of years, we have even built a shared mental model of the universe itself, one which exists inside our brains and which mirrors in imperfect but ever improving detail the cosmos about us. You have been taking part in this process, absorbing and interpreting a version of our cosmic model, as you read this book.

The universe is vast and full of wonders, and the more we have learned about it the more beautiful and amazing it has come to seem. But all our discoveries have reaffirmed something that we have always known: that the Earth and its human passengers are an integral part of the cosmos, intimately connected to its past, present and future. And we are rather amazing too.

Acknowledgements

Everything in the universe is interconnected and this book is no exception. Writing it would not have been possible without the advice, support and encouragement of a huge number of people, including but not limited to: the astronomy team at the Royal Observatory Greenwich and various other colleagues across Royal Museums Greenwich, the team at Quercus, the patient staff at the various cafés where I wrote most of it, and the worldwide community of scientists who continue to make astonishing and inspiring discoveries about the universe in which we find ourselves. Most of all, of course, the friends and family who make this planet such a great place to live.

Index

dark matter 3, 11–12, 256–7, 342–3, 344, 347
daughter elements 14, 301
Dawn spacecraft 70–2, 295, 327, 360
De Revolutionibus orbium coelestium (Copernicus) 221, 227
Deep Impact 69
Deep Impact probe 216
Definition of a Planet in the Solar System 204
degenerate matter 22, 33
Delambre, Jean Baptiste Joseph 243
Denmark 66, 321
deuterium 11, 62
Devonian period 185
*Dialogue Concerning the Two Chief World
Systems* (Galileo) 235–6
diamond 27
dielectric material 42
differential heating 125, 133
digital cameras 356
dinosaurs 69, 193, 212–15, 217, 255
'dirty snowballs' 60
distribution of stars 250–3
diurnal animals 277
diurnal tides 170
DNA (Dioxyribonucleic Acid) 28, 121, 139
Donati's Comet 68
Donne, John 234
Doppler shift 254
Duke Humphrey's Tower, Greenwich 47
dust, cosmic 23, 24, 26–9, 37, 54, 58, 60, 64,
70, 73, 208, 209, 252, 253, 265–7, 301, 316,
317, 321, 331, 347
dwarf galaxies 332
dwarf planets 56, 57, 71, 105, 106, 199–200,
204–7
Dyce, William 68
dynamos 284, 290, 296, 337

$E=mc^2$ 17, 116–17, 126
Earth
angular momentum 265–6, 324
atmosphere 2, 3, 7–8, 15, 20, 40, 52, 55, 65,
68, 75–6, 83, 88, 90, 96, 121–6, 128, 130,
132–4, 137–40, 150–1, 159, 162, 192–3,
208–11, 214, 223, 239, 240, 272, 274,
277–8, 280–2, 284–5, 292–3, 301, 307,
314–15, 318, 321, 333–7, 351, 356–7
axial rotation 114, 125, 184, 185–6, 222–3,
239, 242, 265–86, 324
axial tilt 114, 125, 160–3, 269, 270, 310, 313,
315, 324
carbon dioxide, declining levels of 334–6
climate system 114, 124, 125, 133, 150, 159,
160, 162, 218, 223, 239, 280, 282, 313–15,
317, 324, 325
collision with Theia 157–60
composition of 14, 15–16
core 20, 137, 158, 284, 290–3, 297, 306, 337
Coriolis effect 280–5
crust 52, 90, 158–9, 174, 292, 297, 307
ecosystems 103, 212, 215, 315, 318, 351
equator 46, 125, 133, 174, 272, 273, 274,
275–7, 280, 281, 283
geomagnetic field 137, 138, 140, 142–4, 151,
284–5, 290
gravity 173, 180–1, 186–7, 293

and Habitable Zone 75, 79, 337–9
impacts 208–17, 292–4, 299, 327
ionosphere 121
magnetic field 290, 296, 306, 337
mantle 52, 287–8, 291, 301
Milankovitch cycles 313–15, 324
nutation 163, 242, 243
oceans 1, 2, 4, 7–9, 15, 20, 53, 60, 62–3, 65,
70, 74, 75, 94, 106, 112, 124–6, 134, 159,
162, 163, 168–78, 183–6, 188, 195, 218,
239, 273, 278, 280–3, 307, 314, 324–5,
336–7
orbit 178, 185, 239, 222, 223, 269–71, 310,
313, 315, 325, 338
ozone layer 121, 151, 284, 318, 335
plate tectonics 52, 90, 159–60, 185, 292,
307, 337
polar zones 76, 102, 125, 133, 137, 138, 160,
275, 282, 314–15
primordial heat 300, 306
rotation 114, 125, 184, 185–6, 222–3, 239,
242, 265–86, 324
rotational speed 272–3, 276, 285
seasons 2, 114, 125, 154, 160, 163, 184, 269,
308
shape of 275
temperate zones 133, 160, 282
temperature 76, 79, 125, 277–8, 333–4, 336
tidal locking 180–1, 183, 187–8
tropics 125, 160, 280, 281, 282
water, origins of 52–74
weather 2, 112, 114, 124, 125, 133, 136, 274,
282–3
earthquakes 287–8, 291
Earthrise images 168
Echo, Project 40, 49
eclipses 184, 187
ecosystems 103, 212, 215, 315, 318, 351
Ecuador 275
Egypt 83–6, 87–9, 95, 298, 308
Einstein, Albert 17, 45, 116–17, 126, 241, 261,
345
El Niño 2
electricity 127, 131–4, 138, 142–6, 186, 284,
302
electromagnetic radiation 12, 13, 17, 18, 34, 38,
43, 115, 116, 117, 127, 129, 135, 256, 320,
322, 330
electromagnetism 346, 362
electron micrographs 92
electrons 10, 11, 12, 14, 18, 22, 33, 116, 118,
127, 135, 186, 196, 340
Elephant Moraine meteorite 86
Elizabeth I, queen of England and Ireland 47
elliptical galaxies 194, 331
elliptical orbits 67, 177, 178, 199, 202, 229,
237, 239, 270, 271, 313
Elson, Rebecca 343
Enceladus 104–5, 106, 109, 191
energy 126–35, 138
Enlightenment, Age of 243, 354
epicycles 226, 228
equator 46, 125, 133, 174, 272, 273, 274,
275–7, 280, 281, 283
equatorial metres 277
equinoxes 226

of planets 55, 57, 59, 313, 325–7
of Saturn 192
of stars 16–19, 21–2, 31–4, 35, 196, 304–5, 312, 324, 333, 340
of Sun 192, 204
and tidal locking 107
and tides 2, 169, 172–8, 180, 186–7
of Venus 80
of waterworlds 106
Great Attractor 259, 264
Great Bear 316
Great Comet (1577) 66–7
Great Lakes, North America 111
Great Moon Hoax (1835) 77–8
Greece 65, 147, 202, 222–5, 246, 309
greenhouse gases and effect 80, 81, 106, 126, 130, 149, 159–60, 277, 279, 315, 334, 336
Greenland 298
Greenwich, England 46–51, 149, 154, 168–9, 170, 243, 244
Greenwich Mean Time (GMT) 46–7, 51, 58, 100, 244, 270–1
Greenwich Time Ball 271
Gregorian calendar 154, 228
Gregory XIII, Pope 227, 236
guest stars 31, 320
Guiana Space Centre 273
Gulf Stream 126
gyres 282–3, 285

Habitable Zone 75, 79, 81, 124, 337–9
haematite 82
The Hague, Netherlands 231
Hale–Bopp comet 63, 72
Halley, Edmond 67, 202, 238, 243
Halley's Comet 30, 63, 67, 72
Hardy, Thomas 68
Harold II, king of England 66
Hart, Lorenz 164
Hartley 2 comet 72
Haumea 56, 205
Hawaiian 259
Hawking, Stephen 347
healthcare 357
heat loss 300
heavy elements 11, 15–16, 23, 25, 29, 31, 34
heavy water 11, 62–3, 72
heliocentric model 220–4, 226–9, 234–44, 247–9, 354
heliopause 361
heliosphere 135–6
helium 11, 13, 15, 16, 17, 18, 19, 21, 22, 25, 32, 116, 117, 134, 194, 301, 304, 333, 338, 339, 340
helium 'ash' 21, 338, 339
Henry VIII, king of England 47
Herschel, John 77–8, 355–6
Herschel, William 77, 201, 202, 251–2, 340
Herschel Space Observatory 71
Hevelius, Johannes 149
Hipparcos mission 316
Hiroshima, Japan 211, 212, 221
Hirst, Damien 361
Hodgson, Richard 143
Holmdel, New Jersey, United States 39
Holmdel Horn Antenna 41, 44, 50

Hot Jupiters 194, 267, 325, 326
Hubble, Edwin 52, 260, 261, 262
Hubble Space Telescope 37, 56, 71, 258, 288, 328, 352, 356, 357
human body 6–8, 14, 15–16, 17, 19, 38, 115, 130, 139–40
human brain 363
Humboldt Current 283
Hume, William Fraser 83–4, 86
hurricanes 283
Huygens probe 108–9, 110, 360
Hyakutake 63, 72
Hydro-Québec 143
hydrocarbons 108, 351
hydroelectric power 133
hydrogen 10–11, 13–19, 21, 25, 32, 35, 53, 55, 62, 80, 116–17, 134, 194, 301, 304, 324, 328, 333, 338, 339
hydrogen sulphide 27
hydrothermal vents 103

ice 52, 53, 55–65, 70, 71, 73, 98, 102, 103, 104, 105, 106, 110, 190, 193, 200
Ice Ages 150–1, 163, 255, 314
ice caps 76, 162
ice giants 203
ice sheets 314–15
IK Pegasi 323–4
impact winter 214
impacts 58, 61–2, 71, 73, 102, 179, 182, 207–17, 326, 327
Imperial Porcelain Works, Vienna 296
In the Days of the Comet (Wells) 69
India 86, 189, 281, 309
Indian Ocean 281
Industrial Revolution 129, 354
infrared radiation 18, 24, 29, 115, 120, 121, 124–6, 127, 133, 278
infrared telescopes 321
interglacial period 314
International Astronomical Union (IAU) 199, 204
International Space Station (ISS) 45, 140, 141, 273, 278–9, 359
Internet 354
Inuit 298–9
inverse square law 172
Io 155, 190–1
iodine 35
ionosphere 121
Iran 79
iridium 213
iron 15, 20, 31, 32, 33, 36, 284, 290–1, 293, 294, 295, 296, 297, 301, 305, 324
Iron Age 297
iron crystals 291
iron oxide 297
Islam 309
Islamic calendar 154
ISON comet 192
isotopes 62, 70
jet stream 150
John Paul II, Pope 238
Joshua 220
Juno 201
Jupiter 69, 200, 201, 203, 204, 207, 225, 227, 293

atmosphere 193
composition of 15, 193
Galileo probe 94
gravity 57, 191, 192, 326, 327
moons 53, 55, 101–6, 155, 167, 190–1, 202, 232–4, 237, 338, 362; *see also* Callisto, Europa, Ganymede, Io
orbit 227
Shoemaker–Levy 9 collision 61, 192
Jurassic period 213

Kazakhstan 273
Kepler, Johannes 66–7, 172, 228–30, 236, 237, 238, 243, 264, 309, 354
Kérouaille, Louise de, duchess of Portsmouth 48, 49
Kourou, French Guiana 273
Kraken Mare, Titan 111
Kuiper Belt 56, 59, 69, 72, 200, 205, 206, 207, 299, 326

Lagrangian point 157
Lake Windermere, England 117
Laniakea Supercluster 259, 262
lapis lazuli 298
Large Hadron Collider (LHC) 139
Large Magellanic Cloud 322
Lascaux Caves, France 255
laser eye surgery 357
Late Heavy Bombardment theory 62
latitude 46, 48
lava 76, 78, 179, 180, 182, 190
lava planets 107
laws of planetary motion 229, 238
lead 35
leap seconds 186
Lemaître, Georges 260, 261
'Let There Always Be Light' (Elson) 343
Life of Galileo (Brecht) 221
Ligeia Mare, Titan 111
light 12, 13, 18, 45, 58, 113, 116–24, 127, 131, 138
aberration of 240–3
speed of 45, 58, 120, 242, 345, 361
visible light 122–4
light years 246–7
limestone 9
LINEAR comet 72
lithium 11, 13, 15
Little Ice Age 150–1
Local Group 257–9, 261, 262, 263, 264, 331, 346
London, England 47, 150, 231, 277, 352
longitude 46, 48–9
low Earth orbit 272
Lowell, Percival 95, 97
Luna-3 probe 182
lunar distance method of navigation 154–5
lunar laser ranging 188
lunisolar calendars 154

Madrid, Spain 231
Maeshowe, Scotland 114
magma 78, 86, 97, 98, 107, 157–8, 293
magnesium 15, 26
magnetic fields 80, 81, 103, 136, 137, 138, 140–1, 142–4, 147, 148, 151, 155, 284–5, 290, 296–7, 306, 337
magnetism 299
Maine, United States 171–2
Makemake 56, 205
Manhattan Project 221
mantle 287–8, 291, 300, 301
maria 77
Mariner missions 96
Mars 55, 69, 140–1, 193, 200, 201, 202, 204, 207, 225, 293, 294, 352, 359, 360
atmosphere 63, 79–80, 81–2, 96, 284
axial tilt 162
Beagle 2 lander 361
'canals' 95–6
Curiosity rover 361
equator 274
gravity 81, 94, 100
and Habitable Zone 76, 79, 81, 338
life 91–7, 99, 351
mantle 300
Mariner and Viking missions 96
and meteorites 87–95, 97–100
moons 156, 325
Olympus Mons 96
orbit 227, 271
polar ice caps 314
Rosetta probe 63
surface 89, 95–7, 193, 351, 361
volcanism 87, 300
water 52, 76, 79–80, 81–2, 88, 91–2, 95, 97, 98, 99–101
Mars Atmosphere and Volatile EvolutioN mission (MAVEN) probe 81–2
Mars Science Laboratory (MSL) 101
Mathematical Principles of Natural Philosophy (Newton) 237
Maunder, Walter and Annie 149
Maunder Minimum 149–50
Maya civilization 309
measurements 276–7
medical scanning procedures 357
Medici family 233, 234
medicine 234, 244, 309
Mediterranean Sea 170
menstrual cycle 165–6
Mercury 72–3, 156, 202, 203, 204, 294, 338
atmosphere 72, 193, 279
core 300
and Jupiter's gravity 327
orbit 194, 279, 325, 327
temperature 73, 124, 279
water 73, 279
mercury 305
Mesoamerica 309
Messier, Charles 30
meteor showers 209
meteorites 70, 72, 83–100, 210, 293–9
Allan Hills 84001 91–2
Cape York 298–9
Chassigny 86
Cranbourne 299
Elephant Moraine 86
Gibeon 298
Nakhla 83–6, 87–9, 95, 97–100
Shergotty 86

meteoritics 210
meteorology 66, 274, 309
meteors 209, 210
methane 56, 58, 105, 108, 110, 111, 278, 314
Metis 325
metres 276–7
Mexico 144, 213
microbes 92–3, 103, 104, 105, 162, 185, 335, 337
microwave radiation 10, 13, 18, 39–45, 256, 262, 344, 362
Milankovitch cycles 313–15, 324
Milky Way 23–4, 27, 34, 35, 73, 193, 251–5, 257, 258, 260, 261, 264, 315–19, 320, 321, 328–31, 340
molecular clouds 54
molecules 6–9, 24, 25–9, 115, 127, 128, 130, 137, 301, 346
    formation of 25–9
Moon 4, 49, 66, 81, 87, 101, 153–89, 202, 218, 219, 225, 232, 351, 358, 359, 360
    angular momentum 265–6, 324
    and animal behaviour 165
    Apollo missions 87, 89, 140, 141, 168, 182, 188, 326, 350, 358
    axial rotation 180, 181, 183, 278
    core 300
    craters 62, 179, 182, 208, 232, 234–5, 237, 278, 326, 360
    and crime 164–5
    cultural influence 163–4
    cycle of phases 153–4, 171, 178, 227, 232, 311
    dark side 179
    far side 179, 182, 189
    and gardening 166–7
    gravity 161, 173–4, 176
    light 113
    lunar days 278
    magnetic field, lack of 140–1
    and menstrual cycle 165–6
    meteorites 90
    and navigation 154–5
    and nutation 242
    orbit 49, 171, 174, 177, 183, 187, 311, 324
    origin of 156–60, 267
    retreat from earth 187–9, 324
    rock samples 87, 89, 91
    'seas' 76–9, 179, 180, 182
    size 155
    temperature differences 278
    tidal forces on 180–3, 186–7
    and tidal locking 180–1, 183, 187–8
    and tides 168–78, 324
    water on 278
motion, laws of 238
Mount Everest 96, 196, 275
Mylar 279

Nagasaki, Japan 221
Nakhla meteorite 83–6, 87–9, 97–100
Nama people 298
Naples, Italy 296
Napoleon I, emperor of the French 296
NASA (National Aeronautics and Space Administration)
    Apollo missions 87, 89, 140, 141, 168, 182, 188, 326, 350, 358
    Cassini spacecraft 94, 104, 105, 107, 110, 111
    Dawn spacecraft 70–2, 295, 327, 360
    Deep Impact probe 216
    'Follow the Water' 99
    Galileo probe 94, 101, 103, 360
    Mariner missions 96
    Mars Atmosphere and Volatile EvolutioN mission (MAVEN) probe 81–2
    Mars Science Laboratory (MSL) 101
    NEAR/Shoemaker probe 360
    New Horizons probe 57–9, 71, 146, 199, 206–7, 360, 361
    Pioneer probes 102, 108, 146, 206, 361
    on solar storms 146
    Viking mission 96
    Voyager probes 102, 108, 146, 206, 361
natural gas 129, 131
natural nuclear fission reactor 303
navigation 353–4
neap tides 171, 172, 177
NEAR/Shoemaker probe 360
nebulae 3, 30–1, 35, 37, 54, 70, 301, 316, 339–40
    Crab Nebula 30–1, 35
    Orion Nebula 37, 54
neon 15, 21, 32
neon signs 138
Neptune 55, 57, 193, 199, 202, 204, 207, 267, 325, 326
Netherlands 151, 231
neutral atoms 12–13
neutrino detectors 321–2
neutrinos 33, 34, 322
neutron stars 3, 33–4, 36, 196–8, 305, 347
neutrons 10, 11, 14, 33, 62, 196
New Berlin Observatory 149
New Guinea 144
New Horizons probe 57–9, 71, 146, 199, 206–7, 360, 361
New York, United States 41, 277
New York Sun, The 77–8
Newton, Isaac 67, 122, 172–3, 175, 237, 238, 241, 249–50, 253, 260, 311
nickel 15, 32, 33, 284, 290–1, 293, 294, 295, 296, 297, 301
Nile River 83, 95, 308
nitrates 318
nitrogen 7, 15, 16, 20, 25, 38, 56, 58, 108, 122, 138, 301
nitrogen oxide 318
nitrous oxide 27
Nobel Prize 44
North America 144, 314
North Sea 168
North Wales 2
Northern Lights 138
Nova Scotia, Canada 170
nuclear batteries 302
nuclear fission 134, 302, 303–4
nuclear fusion 17, 21, 22, 31–3, 35, 37–8, 116–19, 134–5, 305, 312, 333, 340
nuclear power 1, 11, 302, 306
nuclei 10, 11, 12, 14, 17, 18, 32, 33, 116, 118, 134, 135, 138, 301–6, 333, 346, 347

nucleosynthesis 13, 17
nutation 163, 242, 243

oblate spheroids 275
Ocean of Storms (Oceanus Procellarum) 77
oceans
   Earth 1, 2, 4, 7–9, 15, 20, 53, 60, 62–3, 65,
      70, 74, 75, 94, 106, 112, 124–6, 134, 159,
      162, 163, 188, 195, 218, 239, 273, 278,
      280–3, 307, 314, 324–5, 336–7
   Enceladus 104–5
   Europa 102–4, 190, 191
   lava oceans 107
   Mars 81–2, 91
   Ocean Planets 106
   tides 168–78, 183–6, 188, 195, 324
   Titan 107–12
   Venus 80–1, 160
oil 129, 131, 335
oil pipelines 142
Old Testament 220
olivine 54
Olympus Mons 96
On the Revolutions of the Heavenly Spheres
   (Copernicus) 221, 227
Oort Cloud 56, 60, 69, 72, 193, 200, 299,
   317
orbital day 268–9
Orcus 200
'ordinary matter' 11
Ordnance Survey 244
Ordovician period 318
'organic' molecules 28
Orion 316, 323
Orion Arm 253, 317–19
Orion Nebula 37
Outer Space Treaty (1967) 359
oxygen 8, 14, 15, 16, 17, 20, 21, 22, 25, 32, 38,
   52, 53, 55, 122, 127, 128, 138, 297, 301,
   339, 340
ozone layer 121, 151, 284, 318

Pacific Ocean 2
Palaeogene period 213
palaeontology 90–1, 185, 319
Pallas 201, 294
palm oil 132
Palomar Observatory, San Diego 200
parachutes 274
parakeets 50
parallax 223–4, 240, 242, 244–5, 251
Paris, France 231
Paris Guns 283
Paris Observatory 47, 149
Parsons, William, Earl of Rosse 30–1
Pegwell Bay, Kent – a Recollection of October
   5th 1858 (Dyce) 68
Penzias, Arno 39–44, 51
Petrie, William Finders 298
Philae lander 64, 69, 360
Philosophiae Naturalis Principia Mathematica
   (Newton) 237
Phobos 325
Phoebe 267
phosphorus 16
photography 355–6

photons 12, 13, 17, 18, 31–2, 44, 45, 117–24,
   131, 135, 138, 150, 322
photosphere 120, 144, 147, 148
photosynthesis 127–32, 167, 214, 332, 334–5
photovoltaic effect 131
pigeons 41–2, 44
'Pillars of Creation' 37
Pioneer probes 102, 108, 146, 206, 361
Pius XII, Pope 238
Placentia Palace, Greenwich 47
planetary migration 325–7
planetary nebulae 23, 339–40
planets
   axial rotation 267–8, 271
   classification of 201–7
   collision of 327–8
   composition of 15
   dwarf planets 56, 57, 71, 105, 106, 199–200,
      204–7
   formation of 12, 13, 24–5, 28–9, 37–8, 73, 85
   gravity 313, 325–7
   Hot Jupiters 194, 267, 325, 326
   temperature differences 279
plants 127–8, 132, 166–7, 214, 332, 334–5
plasma 33, 116, 322
plate tectonics 52, 90, 159–60, 185, 292, 307,
   337
platinum 36, 305
Plough 316
Pluto 56–9, 135, 146
   atmosphere 56
   composition of 57–8
   geology 58–9, 207, 208
   moons 57, 203, 156, 187
   New Horizons probe 57–9, 71, 146, 199,
      206–7, 360, 361
   oceans 105, 106
   orbit 199
   planetary status 199–200, 201, 203–7
   surface 56
plutonium 137
Poland 149, 221
polar ice caps 76, 162
polycyclic aromatic hydrocarbons 28
pop culture 351
potassium 15, 301, 303
power cables 142
power grids 142, 145
Prague, Czech Republic 66
Pre-Raphaelites 68
prime meridian line 46, 51, 244, 271
Principia (Newton) 172
Project Echo 40, 49
Protestantism 230
protons 10, 11, 14, 33, 347
protoplanets 37, 294, 295, 296–7
Proxima Centauri 56, 224, 246, 329
Psalms 220
Ptolemy, Claudius 224, 226, 227, 228, 247, 264
Punga Mare, Titan 111
pyrimidine 28

quantum mechanics 348
Quaoar 200
quarks 10, 11
quartz 213

Sputnik 1 satellite 358
spyglass 230–1
St Pierre 48, 49
Star of Bethlehem 66
star red giants 338
Starry Messenger (Galileo) 233, 234
stars
  binary 35–6, 323–4, 347
  black holes 34, 36, 196–8, 253, 289, 329–31,
    342, 347–8
  Chinese guest stars 31, 320
  constellations 29, 195, 202, 219, 223, 224,
    240, 253, 255, 266, 308, 315–16
  core 305
  death of 22–3, 34–7
  degenerate matter 22, 33
  distribution of 250–3
  first generation of 18–20
  formation of 19, 23, 24, 28, 29, 37–8, 329
  gravity of 16–19, 21–2, 31–4, 35, 196, 304–5,
    312, 324, 333, 340
  lifespan 20–2
  and molecules, formation of 26
  nested shells 21–2
  neutron stars 3, 33–4, 36, 196–8, 305, 347
  nuclear fusion 17, 21, 22, 31–3, 35, 37–8,
    116–19, 134–5, 305, 312, 333
  red dwarfs 26
  red giants 26, 188
  red supergiants 323
  supernovae 3, 27, 34–7, 139, 196, 139,
    304–6, 317–24, 329, 331, 343–4, 347
  white dwarfs 22–3, 33, 34, 36, 324, 339–40,
    347
Steady State theory 42–3
steam power 129
stellar parallax 223–4, 240, 242, 244–5, 251
Stone Age 297–8
Stonehenge, England 114
storms 126, 133
Stradivari, Antonio 151
stromatolites 185
subduction 292
sugarcane 131
sugars 8, 127
sulphur 15
sulphur dioxide 1
sulphuric acid 81
Sun 4, 46, 48, 59, 60, 61, 113–52, 202, 218–19,
    301, 308, 310–11, 315, 316, 329
  axial rotation 266
  circuit of skies 268–70
  colour of 122–4
  composition of 15, 17
  core 16–17, 20, 338, 339
  coronal mass ejections 136, 142, 144, 148
  death of 22, 23, 329, 339–40, 347
  differential heating 125, 133
  formation of 20, 38, 54, 85, 299, 301
  gravity 16, 175, 176–8, 192, 249–50
  Habitable Zone 75, 79, 81, 124, 337–9
  heliocentric model 220–4, 226–9, 234–44,
    247–9, 354
  heliosphere 135–6
  infrared radiation 124–6, 127, 133
  light 2, 8, 40, 64, 71; *see also* photons

magnetic field 147, 148
molecules 26
  nuclear fusion 17, 22, 29, 116–19, 134–5,
    333, 340
  orbit of Milky Way 254, 257, 315–19
  oscillation 254
  photons 12, 13, 17, 18, 31–2, 44, 45, 117–24,
    131, 135, 138, 150, 322
  photosphere 120, 144, 147, 148
  rotation of 136, 148
  solar flares 148
  solar maximum 148–9
  solar power 126–35
  solar storms 136, 137, 138, 141–6, 148
  solar wind 135–41, 146, 284, 286, 306
  sunspots 144, 147–50
  surface temperature 26
  tides 169, 170, 175, 176–8, 338
  transit of Venus 352
  worship of 114–15
sunburn 121
supercritical fluid 106
superflares 151–2
supermassive black holes 139, 253, 329–31, 347
supernovae 3, 27, 34–7, 139, 196, 139, 304–6,
    317–24, 329, 331, 343–4, 347
swords 298

Tarantula Nebula 322–3
Taurus 30
tectonic plates 52, 90, 159–60, 185, 292, 307,
    337
telecommunications 274
telegraph system 145
telephone networks 142
telescopes 3, 23, 24, 77, 102, 167, 202, 217,
    219, 224, 230–8, 240, 244, 246, 249, 250–2,
    288–9, 340, 356–7, 363
  adaptive optics (AO) systems 357
  Airy transit circle telescope 46
  gamma ray telescopes 321
  Hubble Space Telescope 37, 56, 71, 258, 288,
    328, 352, 356, 357
  infrared telescopes 321
  radio telescopes 189, 288, 321
  spyglass 230–1
  X-ray telescopes 321
television 44–5, 362
Tempel 1 comet 216, 360
Thames River 47, 150, 168, 271
Theia 157
thermodynamics 333
Thomson, G. 296
thorium 301, 303, 305
tidal barrages 186
tidal locking 107, 180–1, 183, 187, 194
tidal power 134
tidal tails 195
tides, tidal forces 168–89, 190–8, 324, 325
  and black holes 196–8
  and comets 192–3
  diurnal tides 170
  and evolution 198
  and galaxies 194–5
  and Hot Jupiters 194
  and Jupiter's moons 190–1